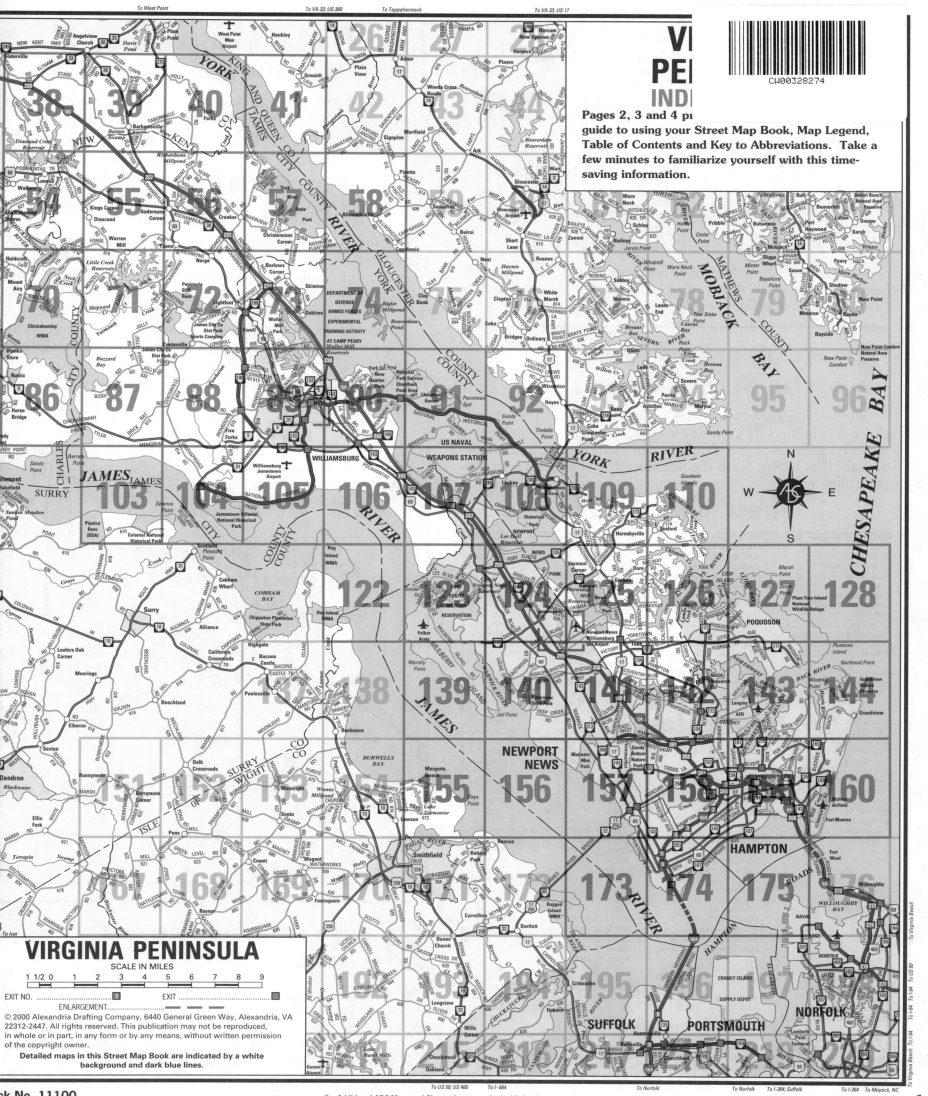

VIRGINIA PENINSULA

SCALE IN MILES

VIRGINIA PENINSULA
INDEX

Pages 2, 3 and 4 pr... guide to using your Street Map Book, Map Legend, Table of Contents and Key to Abbreviations. Take a few minutes to familiarize yourself with this time-saving information.

CW00328274

ck No. 11100

For Additional ADC Maps and Charts please see the inside back cover.

1

A Step-By-Step Guide To Using Your Street Map Book

Thank You For Using The ADC Street Map Book

After you use the Street Map Book once or twice, you won't need any help at all. And you may not need it now. But just in case, we've included this helpful guide for your convenience.

1. It all starts with the Street Index.

Let's say you want to go to the 1500 block of Jamestown Road, which is in James City County. First, you would turn to the Street Index on page 68 of your book and find Jamestown Road in the alphabetical listing. Note the numbers you see listed after the street name.

Street Index Page 68

Holloway Rd 140-C5 NN; 127-A9 Pq	Huntington Ave 157-J10; 173-K1; 174-A3 NN	Jacinth Cir 124-C10 NN	John Smith D
Holly Bk Dr 104-C1 J	Huntington Dr 88-K9 J	Jack Shaver Dr 123-J4 NN	John Smith T
Holly Dr 157-D4 NN	Huntington Rd 159-F7; HE-F8 H	Jacklyn Cir 157-K1 H	John Twine 1
Holly Forks Rd 39-G10 J; 40-A3, D6, D7 NK	Hunts Neck Rd 126-F6 Pq	Jacks Pl 124-C4 NN	John Tyler M CC; 87-B8,
Holly Grove 104-J2 J	Huntshire La 126-K10; 127-A10 Pq	Jackson Ave 123-H6 F	89-A10, B9
Holly Hills Dr 89-D9 W	Huntstree Pl 140-H4 NN	Jackson Cir 141-K1 Y	John Vaughn
Holly La 54-G10 J	Hurley Ave 157-F5 NN	Jackson Ct 142-A4 Y	J
Holly Pt Rd 86-J9 CC; 126-C5 Y	Hurst Dr 159-J3 H	Jackson Dr 90-C8 J	John Wickam
Holly Rd 105-A2 J; 143-C8 L	Hurst St 89-B10 J	Jackson Rd 157-H4 NN	John Wythe D
Holly Ridge La 104-J2 J	Huskie Dr 142-B4 Y	Jackson St 106-K5; 107-A4 J	Johns Landin
Holly St 159-G5; HE-G2 H; 143-B9 L; 143-C1 Pq	Hustings La 124-D9 NN	Jacobs La 140-K10; 156-J1 NN	Johnson Dr 1
Hollyberry St 159-C10 H	Huxley Pl 141-B7 NN	Jacobs Rd 90-D10 J	Johnson La 1
Hollywood Ave 174-H1 H	Hyacinth Cir 124-C10 NN	Jacobs Run 125-H5 Y	Jolama Dr 14
Hollywood Blvd 125-J8 Y	Hyatt Pl 140-J5 NN	Jacqueline Dr 92-J6 G	Jolly Pond Ro
Holm Rd 91-J9 Y	Hyde La 141-H2 Y	Jakes La 124-D4 NN	87-G1, J3
Holman Rd 106-C1 J	Hygeia Ave 159-K6 H	James Blair Dr 89-F7; WE-B3 W	Jonadab Dr 1
Holmes Blvd 125-F3 Y		James Bray Dr 88-F1 J	Jonadab Rd 1
Holston La 144-C8 H		James Ct 159-G8; HE-F8 H	Jonas Profit T
Holt Cir 125-K1 Y	**I**	James Dr 92-J7 G; 54-A7 NK; 158-B9 NN; 125-G4 Y	Jonathan Dr 1
Home Pl 159-H8 H		James Landing Rd 156-H1 NN	Jonathan Jct
Homeland St 174-G1 H	Ibis Pl 126-B4 Y	James Longstreet 106-B2 J	Jonathon Ct 1
Homestead Ave 159-B7 H	Ida St 159-B7 H	James River Dr 157-D5 NN	Jones Dr 72--
Homestead Pl 142-A1 Y	Idaho Cir 73-F4 Y	James River La 140-F10 NN	Jones Mill La
Homestead Rd 38-G10, J8; 39-A6 NK	Idlewood Cir 140-J9 NN; 125-H10 Y	James Sq 105-C1 J	Jones Mill Ro
Hondo Ct 143-J10 H	Idlewood Dr 89-G8 W	James Terr 158-D4 H	Jones Rd 159
Honeysuckle Hill 143-K9 H	Ignoble La 110-C6 Y	Jameson Ave 159-B2 H	Jonestown R
Honeysuckle La 123-H9 NN; 141-J5 Y	Ilene Dr 124-B9 NN	Jamestown Ave 158-J9 H	Jonquil Ct 14
	Ilex Ct 158-J2 H	Jamestown Dr 140-K8 NN	Jonquil La 14
Hook Rd 92-K6 G	Ilex Dr 126-A4 Y	Jamestown Rd 89-C10; 104-G3; 105-A2 J; 89-C10, E8; WE-B4 W	Jordan Dr 15
Hoopes Rd 124-D10 NN	Impala Dr 141-E8 NN	Jameswood 104-A1 J	Jordans Jour
Hop Ct 124-F10 NN	Incinerator La 160-C7 FM	Jamie Ct 124-A10 NN	Joseph Lewis
Hope Ct 140-H3 NN	Incubate Rd 159-G2 H	Jan La 110-D10 Y	Joseph Toppi
Hope St 159-H8 H	Independence Dr N 158-J6 H	Jan Rae Cir 107-C7 Y	Josephs Cros
Hopemont Cir 159-H9 H	Independence Dr S 158-J6 H	Jan-Mar Dr 141-B10 NN	Joshua La 15
Hopemont Dr 140-F10 NN	Independence Pl 141-B4 NN	Janet Cir 141-D10 NN	Joshua Way I
Hopkins Ct 142-H1 Pq	Indian Field Rd 92-A10 Y	Janet Dr 158-D8 H	Joshua Way S
Hopkins St 159-J7 H; 157-G6 NN	Indian Cir 107-B6 J	Janice Ct 157-J2 H	Jotank Turn I
Horan Ct 73-C10 J	Indian Path 88-G9 J	Janice Rd 40-B8 NK	Jouett Dr 124
Horizon La 124-H8 NN	Indian Rd 159-F7; HE-E7 H	Janis Dr 109-F7 Y	Joy Dr 158-A
Horn Hawthorn La 105-C2 J	Indian Springs Ct 89-F8; WE-B4 W	Jarvis Pl 158-A7 NN	Joyce Cir 141
Hornes Lake Dr 87-D10, E10; 103-E1 J	Indian Springs Dr 140-J7 NN	Jasmine Cir 124-D10 NN	Joyce Lee Cir
Hornet Cir 157-H8 NN	Indian Springs Rd 89-F8; WE-B4 W	Jasmine Ct 160-A6 H	Joyce Lee Cir
Hornet St 158-G9 H	Indian Summer Dr 142-E5 Y	Jason Dr 90-E8 Y	Joynes Rd 15
Hornsby La 140-K2 NN	Indian Summer La 73-B8 J	Jay Sykes Ct 160-A4 H	Joys Cir 72--
Hornsbyville Rd 109-D9, E10, F9 Y	Indiana La 73-E4 Y	Jaymoore La 159-J5 H	Juanita Dr 15
Horse Pen Rd 140-D7 NN	Indigo Dam Rd 88-K7; 89-A8 J; 156-J1 NN	Jayne Lee Dr 160-C2 H	Jubal Pl 106-
Horse Run Glenn 140-F1 NN	Indigo Terr 89-A7 J	Jean Mar Dr 126-J10; 142-J1 Pq	Judges Ct 12
Horse Shoe Dr 90-E3, E4 Y	Industrial Blvd 56-B8 J	Jean Pl 126-A10 Y	Judith Cir 14
Horse Shoe Rd 70-A3 CC	Industrial Pk Dr 124-B4, C4, D4 NN	Jebs Dr 174-F4 NN	Judy Ct 159-
Horseshoe Ct 124-E9 NN		Jefferson Ave 123-J5 F; 107-J8; 108-A10; 124-B1, E6, H9; 140-J1; 141-A3, D7; 157-F1, J7; 174-B1 NN	Judy Dr 107-
		Jefferson Ct 126-G8 Pq; 142-A4	Jules Cir 141
			Julia Terr 124
			Julian Pl 158
			June Terr 107
			Juniper Ct 10

2. Finding your page and using the grid coordinates.

As you can see, there are four sets of numbers for Jamestown Road, with the sets separated by semi-colons. The first number in each set is the map number, which you'll find on the upper corner of each page.

Jamestown Rd 89-C10; 104-G3; 105-A2 J; 89-C10, E8; WE-B4 W

Together, the letter/number combinations are grid coordinates. You'll see that the letters run across the top and bottom of each map, while the numbers are on one side. The letters following the coordinates are territory codes, which are referred to on page 63 and tell which city or county the item is in. Following the example, first look on map 104, grid coordinate G3 for Jamestown Road. But, as you'll notice, Jamestown Road reaches the edge of map 104 in about the 1600 block. But right where it ends, there's a notation that says "Joins Map 105."

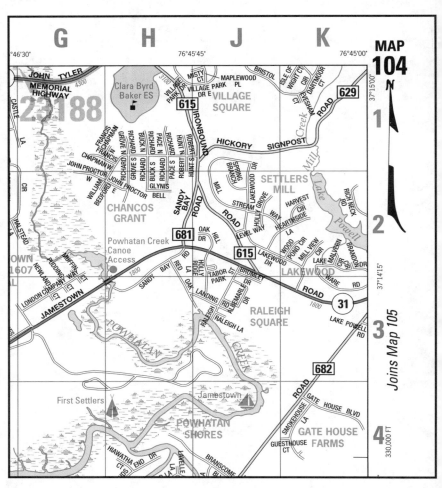

3. So now you've got two choices.

If you want, you can just turn to map 105, and, taking a visual cue from map 104, find Jamestown Road. Or you could go back to the Index where you'll see that Jamestown Road is in grid coordinate A2 on map 105. When you look in that coordinate, you'll see the 1500 block of Jamestown Road. But what's the best way to get there from where you are?

4. Now for the big picture.

You know that the 1500 block of Jamestown Road is on map 105. Now look at the Index to Maps, shown here. Assuming you're in Yorktown you'll see that your current location is in square 108 (square numbers correspond to those of full size maps) and your destination is in square 105. Use the Index to Maps to plot your route from Yorktown to Jamestown Road.

The Index also includes Place Names, Shopping Centers, Schools, Places of Worship and more, all broken down by category. So if you have a name, but not a precise address, these listings will help you find your destination. It's also a good idea to study the Map Legend so you'll be acquainted with the various symbols used on the maps. Knowing them will help you find things more quickly.

Finally, let us hear from you. We like getting ideas for improving the ADC Street Map Books. Write to: ADC, 6440 General Green Way, Alexandria, VA 22312-2447, or fax your comments to 703-750-1591 (Toll-free 1-800-845-9021). Your suggestions are always welcome.

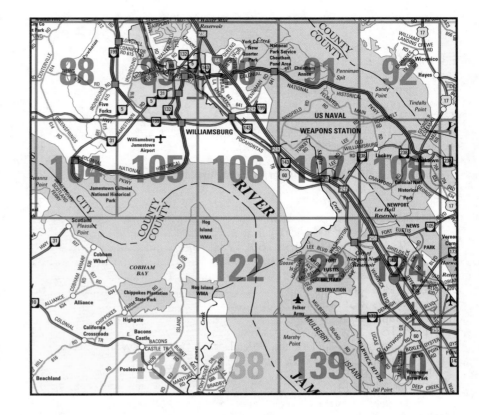

Virginia Peninsula Map Legend

ROADS AND ROUTE SHIELDS

Interstate.. 64

Other Controlled Access Highway..............................

Interchange/Exit Number............................. 255

U.S.; Primary State; Secondary State.......... 60 171 632

U.S. or Primary State Highway..................................

Through Route; Traffic Direction Flow.......... ← BRIARFIELD ROAD

Highway Proposed or Under Construction.......

Street; Block Number................................. WILLCOX NECK RD 9300

Proposed Street..

Trail...

PLACE NAMES

Area Name.. LIGHTFOOT

Subdivision.. WHITTAKER MILL

Business Park...................................... COPELAND IND PK

BOUNDARIES

State.. VIRGINIA NORTH CAROLINA

County.. NEW KENT CO JAMES CITY CO

Incorporated.. HAMPTON

Boundary Definition...............................

Postal Zone/ZIP Code............................. 23666

ZIP Code for Post Office Boxes................. ◯ 23127

OUTDOOR FEATURES

Park, Forest, Recreation or Wildlife Area..........

Community/Recreation Center; Playground....... PG

Campsite; Picnic Area............................

Golf Course or Country Club.....................

Beach; Ski Area.....................................

Marina; Boat Ramp/Launch.......................

Swamp..

Scale 1"= 2000'

1   ½   0   1 Mile

2000 1000 0 2000 4000 6000 8000 Feet

1   ½   0   1 Kilometer

MASS TRANSIT

Railroad; Station................................... —+—S—+—

Park & Ride/Commuter Parking.................. P

GOVERNMENT & PUBLIC FACILITIES

Military, Federal, State or County Property........

Airport..

Hospital.. H Riverside Hospital

Special Property Limits...........................

School; College or University...................

Places of Worship..................................

Cemetery...

Municipal Building................................. 🏛 City Hall

Library; Post Office............................... L P

Fire Station/EMS; Police Station............... F P

Information Center; Parking...................... ? P

Point of Interest; Gate/Barrier.................. ● ⊷

BLACK NUMBERS AND BLUE GRID TICKS INDICATE LATITUDE AND LONGITUDE
BLACK NUMBERS AND BLACK TICKS INSIDE THE NEATLINE INDICATE THE 10,000 FOOT GRID BASED ON THE VIRGINIA COORDINATE SYSTEM
THE HORIZONTAL DATUM REFERENCED IN THIS STREET MAP BOOK IS BASED ON THE NORTH AMERICAN DATUM OF 1983 (NAD 83).

Table of Contents

Key to Abbreviations

Admin .... Administration	Driver Ed .... Driver Education School	Mem .... Memorial	Resv .... Reservation
Al .... Alley	E .... East	Meth .... Methodist	RR .... Railroad
Alt .... Alternate	EMS .... Emergency Medical Services	METRO .... Washington Metropolitan Area	Rte .... Route
AME .... African Methodist Episcopal	Epis .... Episcopal	Transit Authority	S .... South
Apts .... Apartments	ES .... Elementary School	Mgmt .... Management	SC .... South Carolina
Ave .... Avenue	Est .... Estate	MHP .... Mobile Home Park	Sch .... School
Bapt .... Baptist	Ests .... Estates	Mil .... Military	Sec .... Secondary
Batl .... Battlefield	Expwy .... Expressway	Mon .... Monument	SEPTA .... Southeastern Pennsylvania
Bd of Ed .... Board of Education	Ext .... Extension	MS .... Middle School	Transportation Authority
Bk .... Brook	Fed .... Federal	Mt .... Mount	SHA .... State Highway Administration
Bldg .... Building	Frwy .... Freeway	MTA .... Mass Transit Administration	Shop Ctr .... Shopping Center
Bltwy .... Beltway	GA .... Georgia	Mtn .... Mountain	SHS .... Senior High School
Blvd .... Boulevard	GC .... Golf Course/Club	Mun .... Municipal	Soc .... Society
Boro .... Borough	Gdns .... Gardens	Mus .... Museum	Spec .... Special
Br .... Branch/Bridge	Gen .... General	MVA .... Motor Vehicle Administration	Spec Ed .... Special Education
Bus .... Business	Govt .... Government	N .... North	Sq .... Square
By-P .... By-Pass	Hdq .... Headquarters	Nat .... Natural	St .... Saint/State/Street
Cath .... Catholic	Hgts .... Heights	Natl .... National	Sta .... Station
CC .... Country Club	Hist .... Historic/Historical	Nat Res .... Natural Resource	Tech .... Technical
Cem .... Cemetery	Hlth Ctr .... Health Center	Nbhd .... Neighborhood	Terr .... Terrace
Ch .... Church	Hosp .... Hospital	NC .... North Carolina	Tpk .... Turnpike
Chr .... Christian	HOV .... High Occupancy Vehicle	NJ .... New Jersey	Tr .... Trail
Cir .... Circle	HS .... High School	No .... Number	Tr Ct .... Trailer Court
Co .... Company/County	Hwy .... Highway	NRMA .... Natural Resource Management Area	Tr Pk .... Trailer Park
Coll .... College	I .... Interstate	NY .... New York	Trans .... Transportation
Comm .... Community	Ind .... Industrial	Orth .... Orthodox	Trk .... Truck
Condos .... Condominiums	Inst .... Institute/Institution	PA/ Penn .... Pennsylvania	Twnhse .... Townhouse
Cong .... Congregational	Intl .... International	Pent .... Pentecostal	Twp .... Township
Conn .... Connector	IS .... Intermediate School	Pk .... Park	Unitn .... Unitarian
Corp .... Corporate	Is .... Island	Pkwy .... Parkway	Univ .... University
Cr .... Creek	Jct .... Junction	Pl .... Place	US .... Federal Route/United States
Cres .... Crescent	JHS .... Junior High School	PO .... Post Office	UT .... Uptown Map
Cswy .... Causeway	Jr Coll .... Junior College	Prep .... Preparatory	VA .... Virginia
Ct .... Court	JS .... Junior School	Presb .... Presbyterian	VDOT .... Virginia Department of Transportation
Ctr .... Center	La .... Lane	Prim .... Primary	Vet .... Veterans
DC .... District of Columbia	Ldg .... Landing	Prof .... Professional	Vill .... Village
DE .... Delaware	LDS ..... Church of Jesus Christ of Latter Day Saints	Prot .... Protestant	Vis .... Visitors
Dept .... Department	Lp .... Loop	Pt .... Point	Vly .... Valley
Dev .... Development	Luth .... Lutheran	Rd .... Road	Voc .... Vocational
Dist .... District	MARC .... Maryland Rail Commuter	Rec .... Recreation	VoTech .... Vocational/Technical
DMV .... Department of Motor Vehicles	MARTA .... Metropolitan Atlanta Rapid	Ref .... Refuge	VRE .... Virginia Rail Express
DN .... Downtown Map	Transit Authority	Reg .... Regional	W .... West
DOT .... Department of Transportation	MD .... Maryland	Res .... Reservoir	WMA .... Wildlife Management Area
Dr .... Drive	Med Ctr .... Medical Center	Res Sqd .... Rescue Squad	WV .... West Virginia

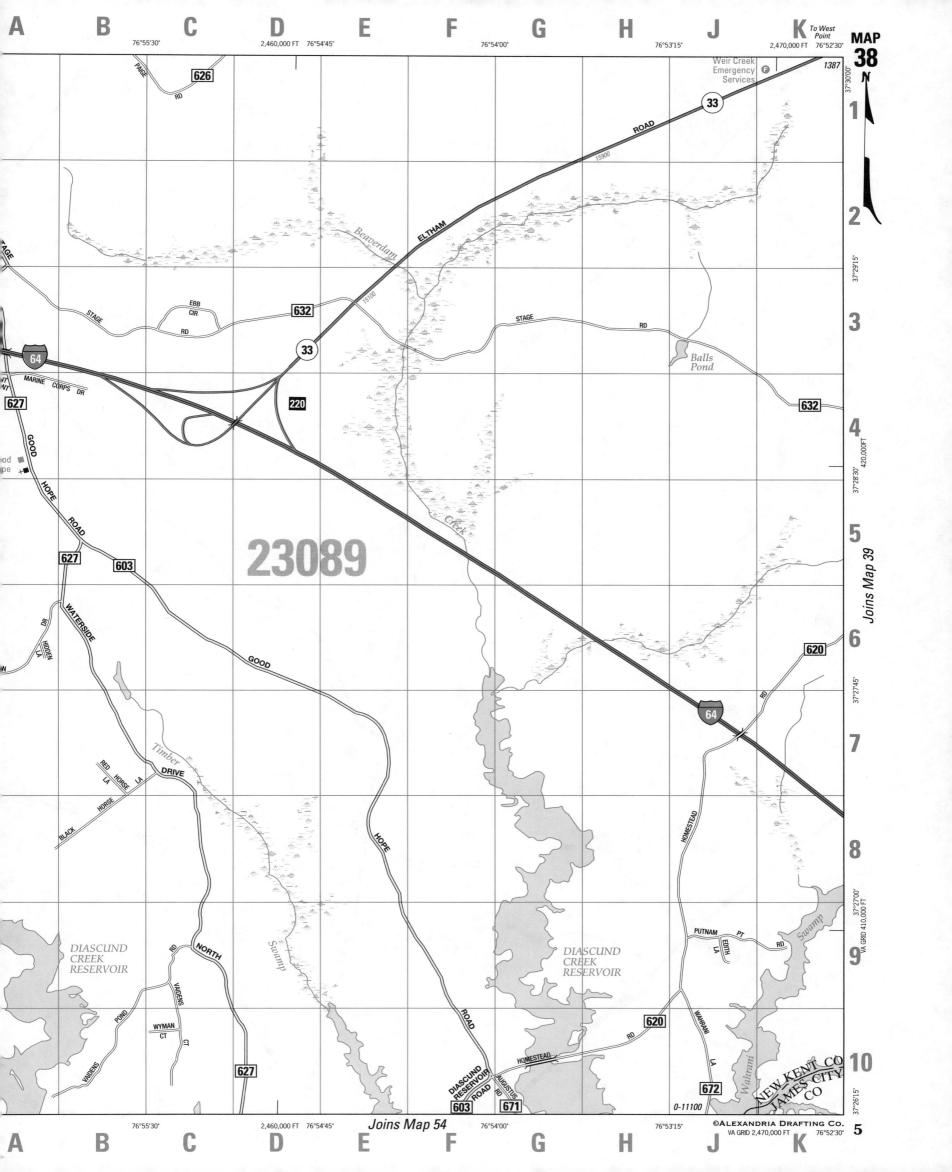

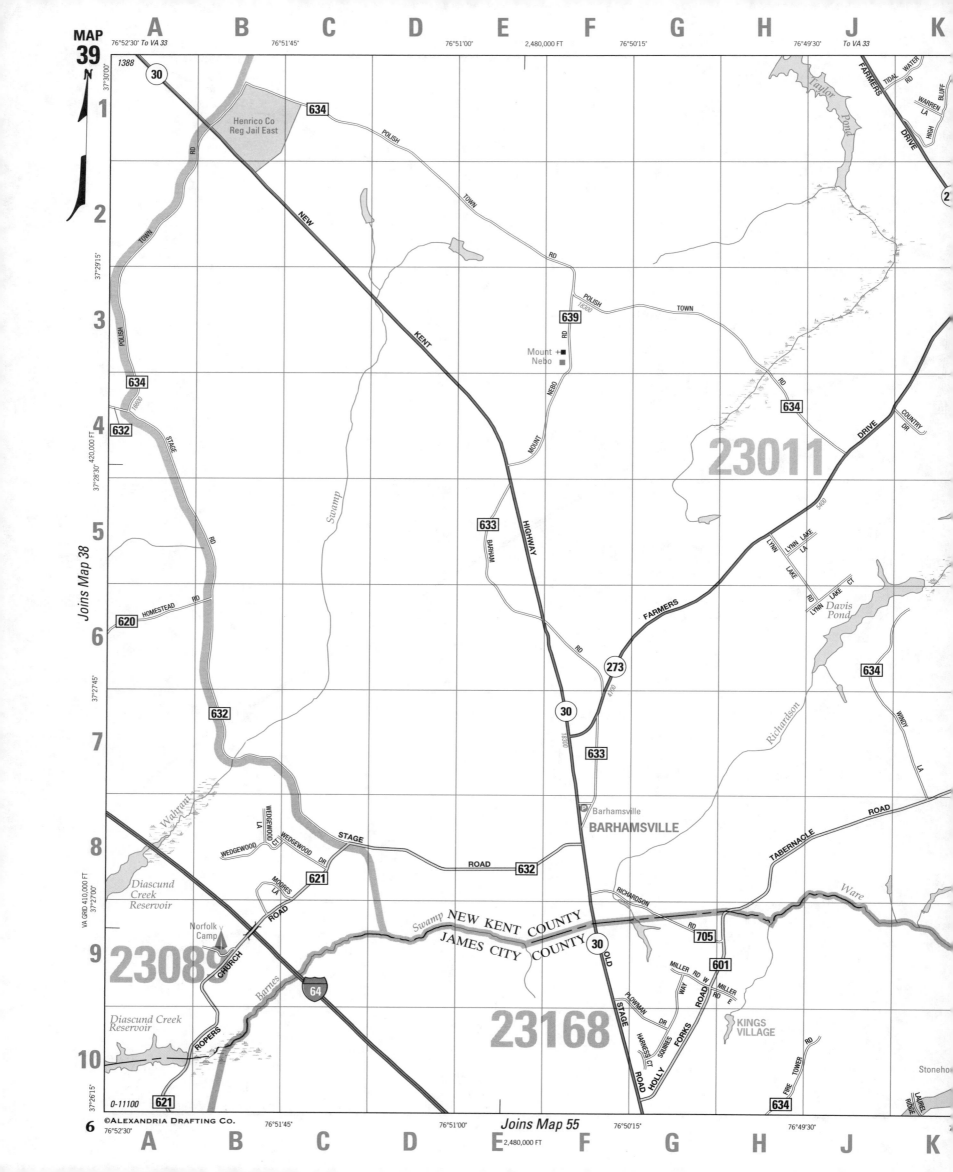

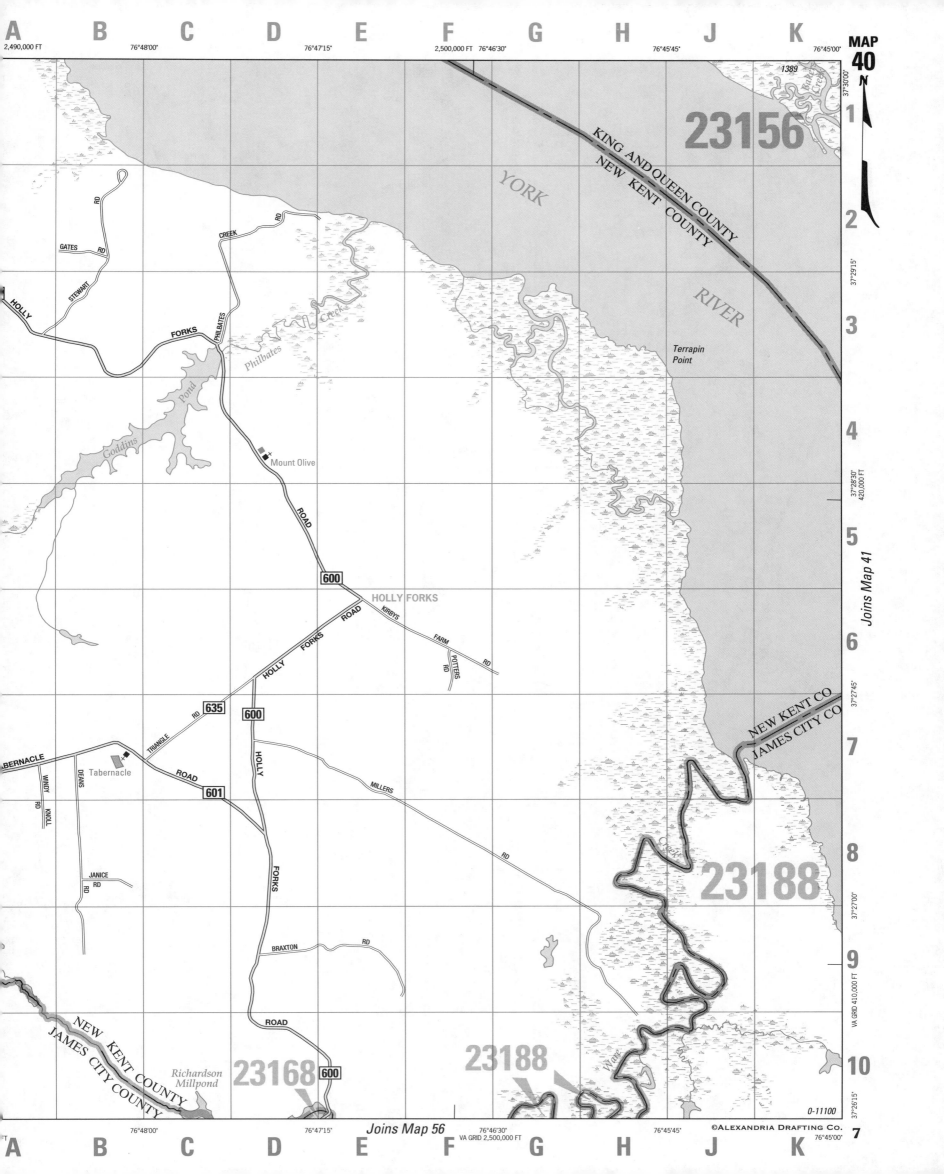

MAP 40
N

A   B   C   D   E   F   G   H   J   K

2,490,000 FT   76°48'00"   76°47'15"   2,500,000 FT   76°46'30"   76°45'45"   76°45'00"

37°30'00"

1389

Buller Creek

23156

YORK

KING AND QUEEN COUNTY
NEW KENT COUNTY

1

2

37°29'15"

RIVER

HOLLY
GATES   RD
STEWART   RD

RD
CREEK   RD

Philbates   Creek

Terrapin
Point

3

37°28'45"

FORKS   PHILBATES

Philbates

Pond

Goddins   Pond

ROAD

Mount Olive

4

37°28'30"
420,000 FT

Joins Map 41

5

600

HOLLY FORKS

KIRBYS

ROAD   FARM   RD

HOLLY   FORKS
POTTERS   RD
RD

6

37°27'45"

RD   635   600

TRIANGLE

NEW KENT CO

JAMES CITY CO

7

BERNACLE

WINDY   RD

DEANS

Tabernacle

ROAD   601

HOLLY

ROAD

MILLERS

RD

Creek

23188

8

37°27'00"

JANICE   RD

FORKS

BRAXTON   RD

9

VA GRID 410,000 FT

ROAD

Water   Creek

10

37°26'15"

NEW KENT COUNTY
JAMES CITY COUNTY

Richardson
Millpond

23168   600

23188

0-11100

A   B   C   D   E   F   G   H   J   K

FT   76°48'00"   76°47'15"   76°46'30"
VA GRID 2,500,000 FT   76°45'45"   76°45'00"

©Alexandria Drafting Co.

7

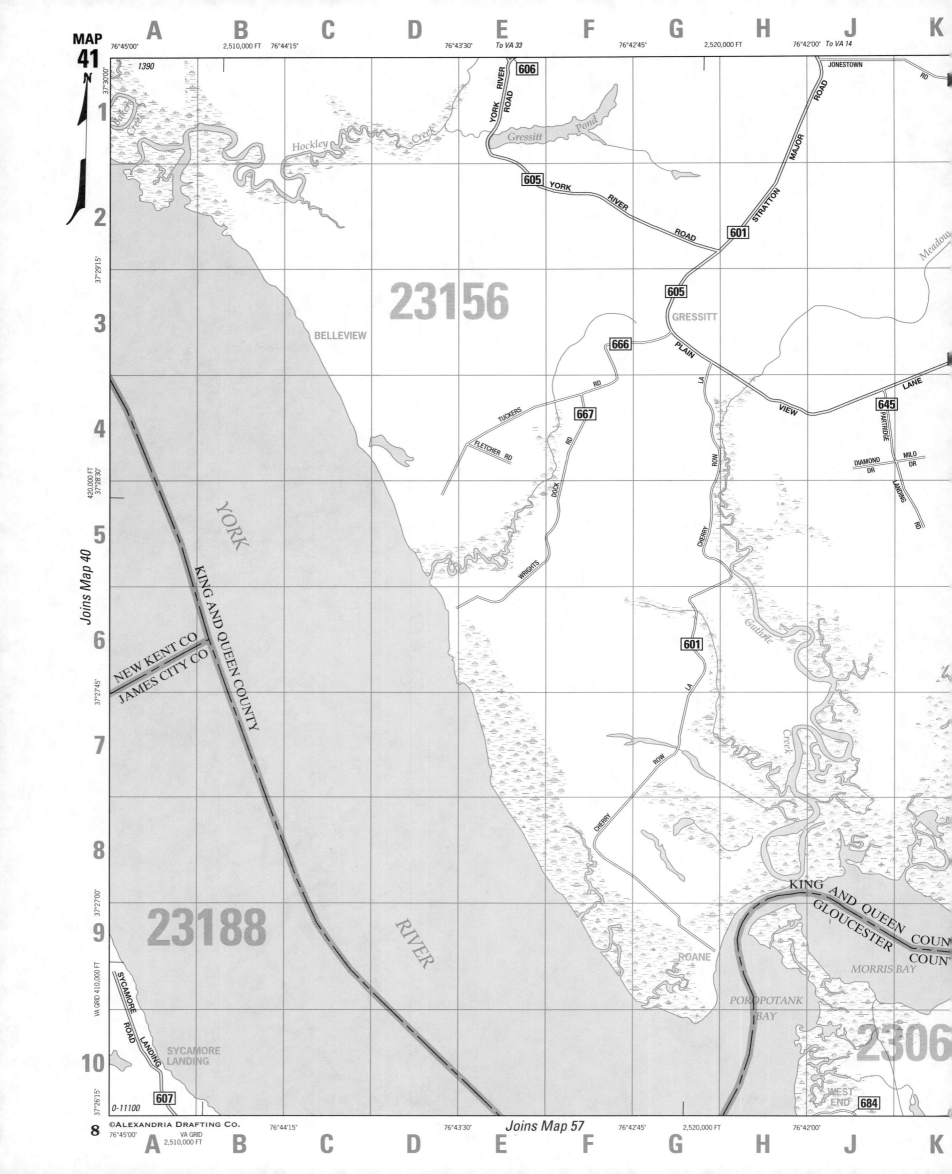

MAP 41

A   B   2,510,000 FT   76°44'15"   C   D   76°43'30"   E   *To VA 33*   F   76°42'45"   G   2,520,000 FT   H   76°42'00"   *To VA 14*   J   K

76°45'00"

37°30'00"

1390

1

Hockley

Creek

Bakers Creek

YORK RIVER ROAD

606

Gressitt Pond

JONESTOWN

RD

2

37°29'15"

605 YORK

RIVER

ROAD

601

STRATTON

MAJOR

ROAD

Meadow

3

23156

BELLEVIEW

605

GRESSITT

666

PLAIN

LANE

4

37°28'30"

420,000 FT

TUCKERS

RD

667

FLETCHER RD

DOCK

RD

CHERRY ROW

LA

VIEW

645

PARTRIDGE

DIAMOND DR

MILO DR

LANDING RD

5

WRIGHTS

Joins Map 40

YORK

KING AND QUEEN COUNTY

6

NEW KENT CO

JAMES CITY CO

601

LA

Gullview

37°27'45"

7

CHERRY ROW

Creek

8

37°27'00"

9

23188

RIVER

ROANE

KING AND QUEEN COUNTY

GLOUCESTER COUNTY

MORRIS BAY

2306

VA GRID 410,000 FT

SYCAMORE

ROAD

LANDING

PAMUNKEY

BAY

10

SYCAMORE LANDING

WEST END

684

0-11100

607

37°26'15"

76°45'00"

VA GRID
2,510,000 FT

76°44'15"

A   B   C   D   76°43'30"   E   F   *Joins Map 57*   76°42'45"   G   2,520,000 FT   H   76°42'00"   J

8

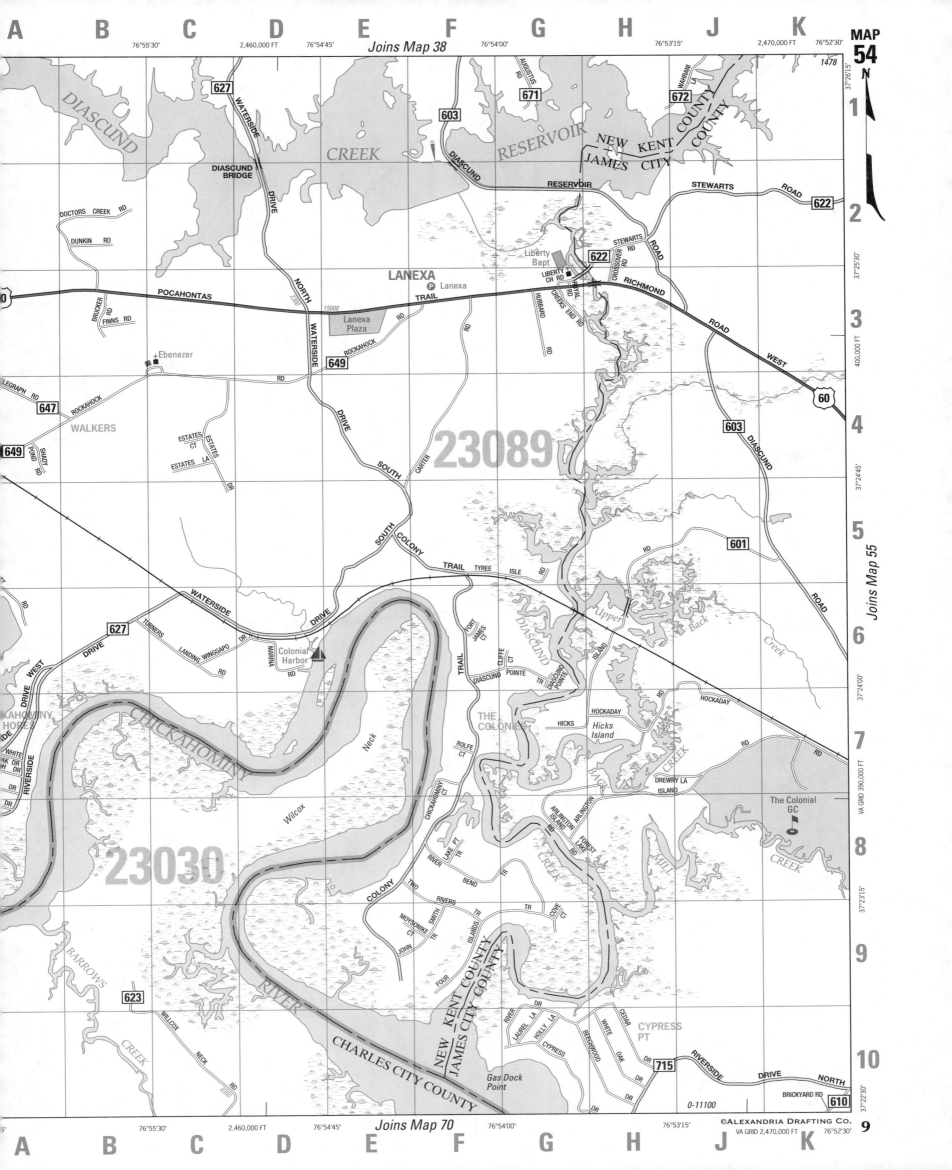

MAP **54**

1478

N

A B C D E F G H J K

76°55'30' | 2,460,000 FT | 76°54'45' | 76°54'00' | 76°53'15' | 2,470,000 FT | 76°52'30'

627

603

DIASCUND

CREEK

RESERVOIR

671

672

WAHRANI LA

NEW KENT COUNTY

JAMES CITY COUNTY

WATERSIDE DRIVE

DIASCUND BRIDGE

AUGUSTUS RD

RESERVOIR

STEWARTS ROAD

622

1

2

DOCTORS CREEK RD

DUNKIN RD

BRUCKER RD

FINNS RD

POCAHONTAS TRAIL

NORTH

2300

LANEXA

P Lanexa

Liberty Bapt

LIBERTY CH RD

622

STEWARTS RD

ROYAL

CROSSOVER RD

RICHMOND

ROAD

37°26'15'

37°25'30'

37°25'00'

+ Ebenezer

15000

Lanexa Plaza

RD

RD

ROCKAHOCK

649

HUBBARD

RD

CREEKS END RD

9500

400,000 FT

3

LEGRAPH RD

647

ROCKAHOCK

WALKERS

RD

WATERSIDE

SOUTH

DRIVE

RD

60

603

DIASCUND

WEST

4

649

SHADY POND RD

ESTATES CT

ESTATES LA

ESTATES DR

CARTER

SOUTH

23089

601

37°24'45'

5

WATERSIDE

627

TURNERS

LANDING

WINGGAPO

DR

RD

SOUTH COLONY TRAIL

DRIVE

Colonial Harbor

MARINA

TYREE ISLE RD

FORT JAMES CT

CLIFFE CT

DIASCUND

DIASCUND POINTE TR

DIASCUND POINTE

Upper

Back

HOCKADAY

HOCKADAY

Creek

37°24'00'

Joins Map 55

6

WEST

DRIVE

CHICKAHOMINY

Neck

Wilcox

THE COLONIES

ROLFE CT

Hicks

HICKS

Hicks Island

RD

DREWRY LA Island

VA GRID 390,000 FT

7

KAHOMINY HORES

WHITE OAK DR

RIVERSIDE

DR

DR

CHICKAHOMINY CT

LAKE PT TR

RIVER BEND

ARLINGTON ISLAND

ARLINGTON RD

FOREST LAKE

MILL

The Colonial GC

CREEK

8

23030

COLONY

TWO RIVERS TR

SMITH TR

TR

COVE CT

RD

RD

CREEK

37°23'15'

9

BARROWS

623

MOYSONIKE CT

JOHN TR

ISLANDS

FOUR

RIVER DR

LAUREL LA

HOLLY LA

CEDAR DR

WHITE OAK

BEECHWOOD DR

CYPRESS PT

715

RIVERSIDE

DRIVE

NORTH

610

BRICKYARD RD

37°23'00'

37°22'30'

10

WILLCOX

NECK

RD

5000

CREEK

RIVER

CHARLES CITY COUNTY

NEW KENT COUNTY

JAMES CITY COUNTY

Gas Dock Point

CYPRESS

0-11100

76°55'30' | 2,460,000 FT | 76°54'45' | 76°54'00' | 76°53'15' | VA GRID 2,470,000 FT | 76°52'30'

A B C D E F G H J K

©ALEXANDRIA DRAFTING CO.

9

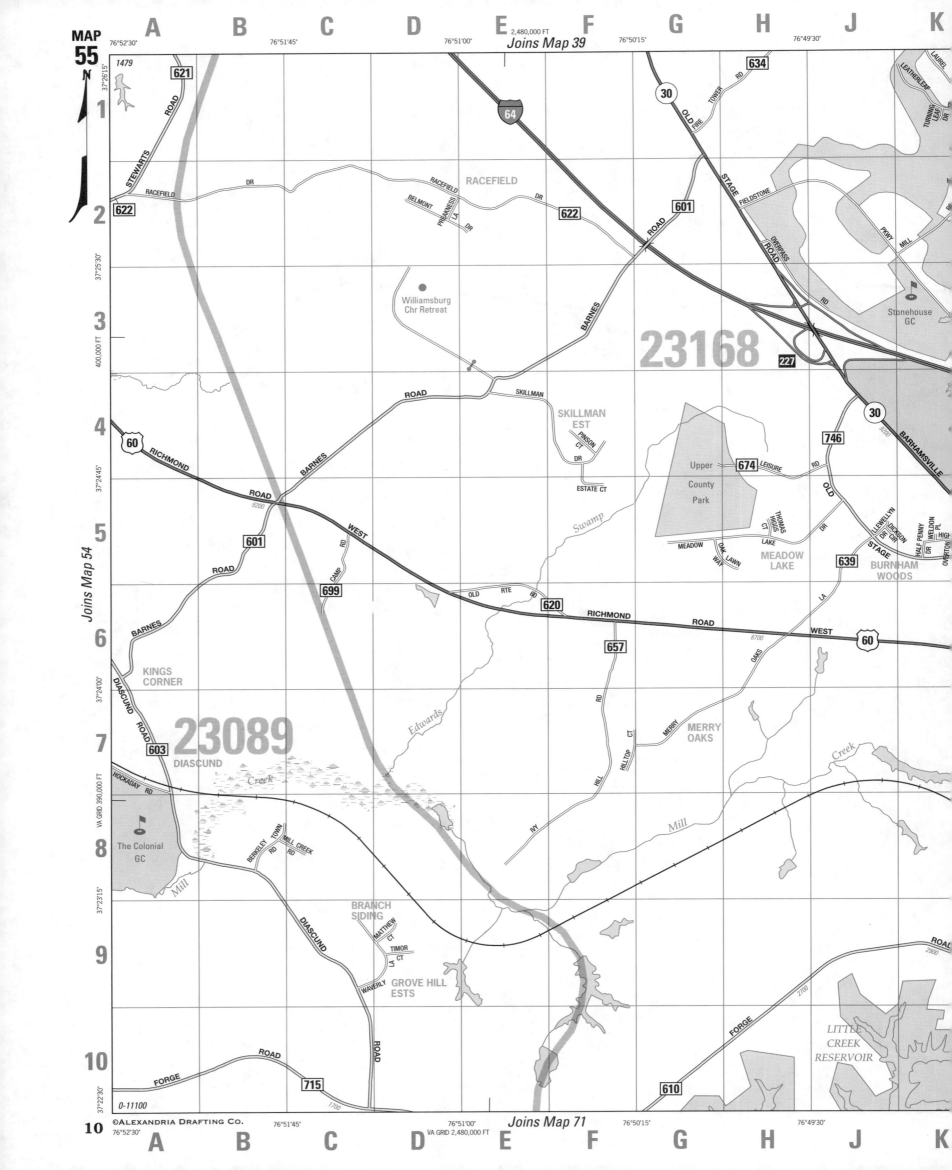

MAP
55

N

1479

STEWARTS ROAD

621

76°52'30"
37°26'15"

634
30
OLD FIRE TOWER RD

76°49'30"

LAUREL
LEATHERLEAF DR
TURNING LEAF DR
TURNING LEAF PL

622
RACEFIELD
DR

76°51'45"

BELMONT DR
PREAKNESS LA
PRIANKESS

RACEFIELD
DR

RACEFIELD

DR

622

BARNES

ROAD

601

STAGE

FIELDSTONE

OVERPASS

RD

PKWY
MILL

227

37°25'30"

Williamsburg
Chr Retreat

23168

Stonehouse
GC

400,000 FT

ROAD

SKILLMAN

SKILLMAN
EST

PINSON
CT

DR

30

BARHAMSVILLE

60
RICHMOND

BARNES

ROAD
9200

ESTATE CT

746

674
LEISURE RD

OLD

Upper
County
Park

THOMAS
HIGGS
CT

LEWELLYN DR CIR
DICKSON CIR

LLEWELLYN DR

HALF PENNY DR
WELDON
OVERTON
HIGH

37°24'45"

37°24'45"

601

WEST

MEADOW

OAK LAWN WAY

MEADOW
LAKE

STAGE

639

BURNHAM
WOODS

Joins Map 54

ROAD

CAMP RD

699

Swamp

OLD   RTE   60

620

RICHMOND   ROAD
8700

WEST

60

LA
OAKS

657

RD

MERRY

MERRY
OAKS

37°24'00"

BARNES

KINGS
CORNER

DIASCUND ROAD

Edwards

HILL

HILLTOP CT

Creek

603

23089

DIASCUND

HOCKADAY RD

Creek

IVY

Mill

VA GRID 390,000 FT

The Colonial
GC

37°23'15"

Mill

BERKELEY RD
TOWN
MILL CREEK RD

DIASCUND

BRANCH
SIDING

MATTHEW CT

TIMOR
LA CT

WAVERLY

GROVE HILL
ESTS

LITTLE
CREEK
RESERVOIR

FORGE ROAD
2700
2900

610

37°22'30"

FORGE

ROAD

ROAD

715
1700

0-11100

76°51'45"

76°51'00"
VA GRID 2,480,000 FT

Joins Map 71

76°50'15"

76°49'30"

76°52'30"

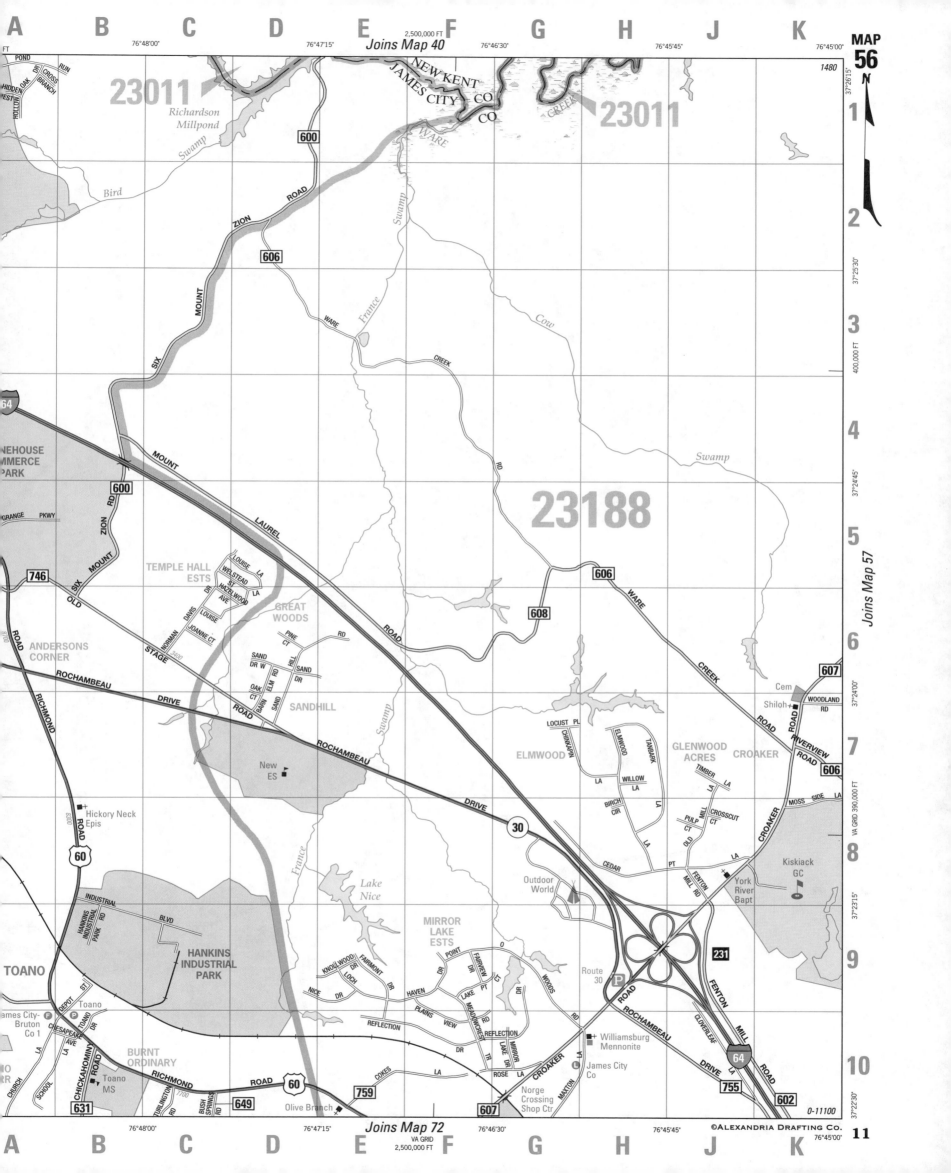

MAP 56

Joins Map 57

1480

N

2,500,000 FT

76°48'00"
76°47'15"
76°46'30"
76°45'45"
76°45'00"

37°2615"
37°25'30"
400,000 FT
37°2445"
37°24'00"
VA GRID 390,000 FT
37°23'15"
37°22'30"

©ALEXANDRIA DRAFTING CO.

0-11100

VA GRID
2,500,000 FT

76°48'00"
76°47'15"
76°46'30"
76°45'45"
76°45'00"

23011
23188
23011

NEW KENT CO
JAMES CITY CO

Richardson Millpond
Bird
Swamp
Ware
Swamp
France
Creek
Cow
Swamp

ZION ROAD
SIX MOUNT
MOUNT
ZION ROAD
SIX MOUNT
OLD STAGE ROAD
LAUREL
ROCHAMBEAU DRIVE
RICHMOND ROAD
ROCHAMBEAU
RD

600
606
600
746
608
606
607
606
30
231
60
755
602
607
759
649
631

NEHOUSE COMMERCE PARK
GRANGE PKWY
ANDERSONS CORNER
TEMPLE HALL ESTS
GREAT WOODS
SANDHILL
New ES
Hickory Neck Epis
Lake Nice
France Swamp
Swamp

LOUISE LA
WELSTEAD ST
HAZLWOOD LA
AVE
DAVIS
LOUISE
JOANNE CT
NORMAN DR
PINE CT
SAND DR W
HILL RD
ELM RD
SAND DR
OAK CT
BARN
SAND
RD

WARE
CREEK
ELMWOOD
LOCUST PL
CHINKAPIN
ELMWOOD
TANBARK
GLENWOOD ACRES
CROAKER
WILLOW LA
LA
BIRCH CIR
TIMBER LA
CROSSCUT CT
PULP CT
OLD
LA
CEDAR
PT
CROAKER ROAD
RIVERVIEW ROAD
WOODLAND RD
Cem
Shiloh
MOSS SIDE LA
Kiskiack GC
York River Bapt
FENTON MILL RD
Outdoor World

MIRROR LAKE ESTS
KNOLLWOOD DR
FAIRMONT DR
LOCH
NICE DR
HAVEN
PLAINS VIEW
REFLECTION
POINT DR
FAIRVIEW DR
LAKE PT
MEADOWCREST DR
WOODS RD
MIRROR LAKE DR
ROSE DR
REFLECTION LA
MAXTON
CROAKER RD
Route 30
Williamsburg Mennonite
James City Co
Norge Crossing Shop Ctr
ROCHAMBEAU ROAD
CLOVERLEAF DR
FENTON MILL ROAD

TOANO
HANKINS INDUSTRIAL PARK
INDUSTRIAL PARK RD
BLVD
BURNT ORDINARY
RICHMOND ROAD
James City-Bruton Co 1
DEPOT ST
Toano
Toano
CHESAPEAKE AVE
CHURCH SCHOOL
CHICKAHOMIN
Toano MS
TURLINGTON
BUSH SPRINGS RD
Olive Branch

8300
8200
7700
3400
7800

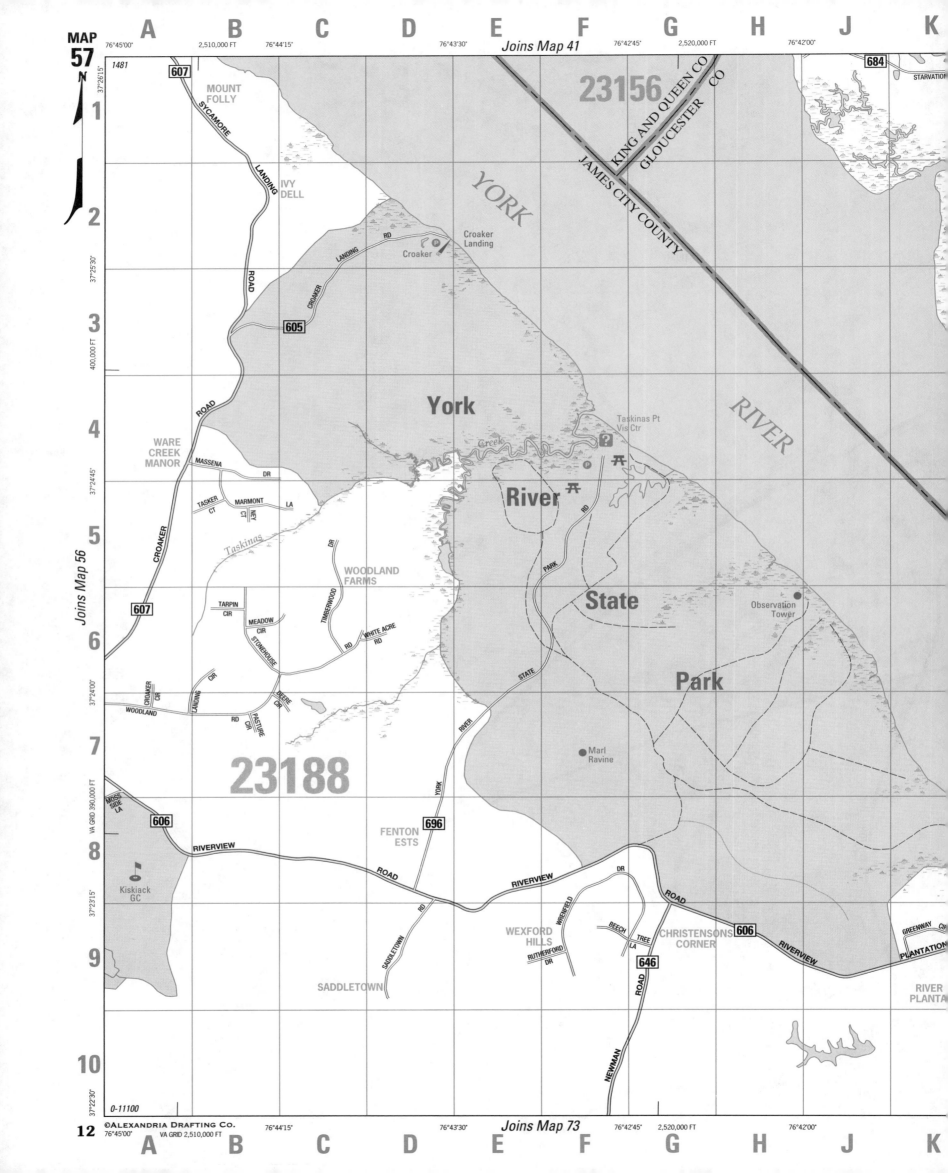

MAP
57

N

A B C D E F G H J K

76°45'00"   2,510,000 FT   76°44'15"   76°43'30"   76°42'45"   2,520,000 FT   76°42'00"

1481

684

607

23156

STARVATION

MOUNT
FOLLY

KING AND QUEEN CO

GLOUCESTER CO

1

IVY
DELL

SYCAMORE
LANDING

JAMES CITY COUNTY

YORK

RD

Croaker
Landing

2

37°26'15"

ROAD

CROAKER
LANDING

605

P

Croaker

37°25'30"

3

400,000 FT

York

RIVER

ROAD

Taskinas Pt
Vis Ctr

4

37°24'45"

WARE
CREEK
MANOR

MASSENA

DR

Creek

?

River

P

37°24'45"

Joins Map 56

CROAKER

TASKER
CT

MARMONT
CT

NEY

LA

5

Taskinas

PARK

607

DR

TIMBERWOOD

Woodland
Farms

STATE

Observation
Tower

6

TARPIN
CIR

MEADOW
CIR

STONEHOUSE

RD

WHITE
ACRE
RD

Park

37°24'00"

CROAKER
CIR

CIR

DEERE
CIR

STATE

7

WOODLAND

LANDING

RD

PASTURE
CIR

YORK

Marl
Ravine

23188

37°23'45"

390,000 FT

MOSS
SIDE LA

606

RIVER

8

37°23'15"

Kiskiack
GC

RIVERVIEW

ROAD

FENTON
ESTS

696

RIVERVIEW

DR

ROAD

606

RIVERVIEW

9

SADDLETOWN

RD

WEXFORD
HILLS

WRENFIELD

BEECH
TREE
LA

CHRISTENSONS
CORNER

GREENWAY
CIR

PLANTATION

RUTHERFORD
DR

646

SADDLETOWN

NEWMAN

RIVER
PLANTA

10

37°22'30"

0-11100

12

©Alexandria Drafting Co.

76°45'00"   76°44'15"   76°43'30"   76°42'45"   2,520,000 FT   76°42'00"
VA GRID 2,510,000 FT

A B C D E F G H J K

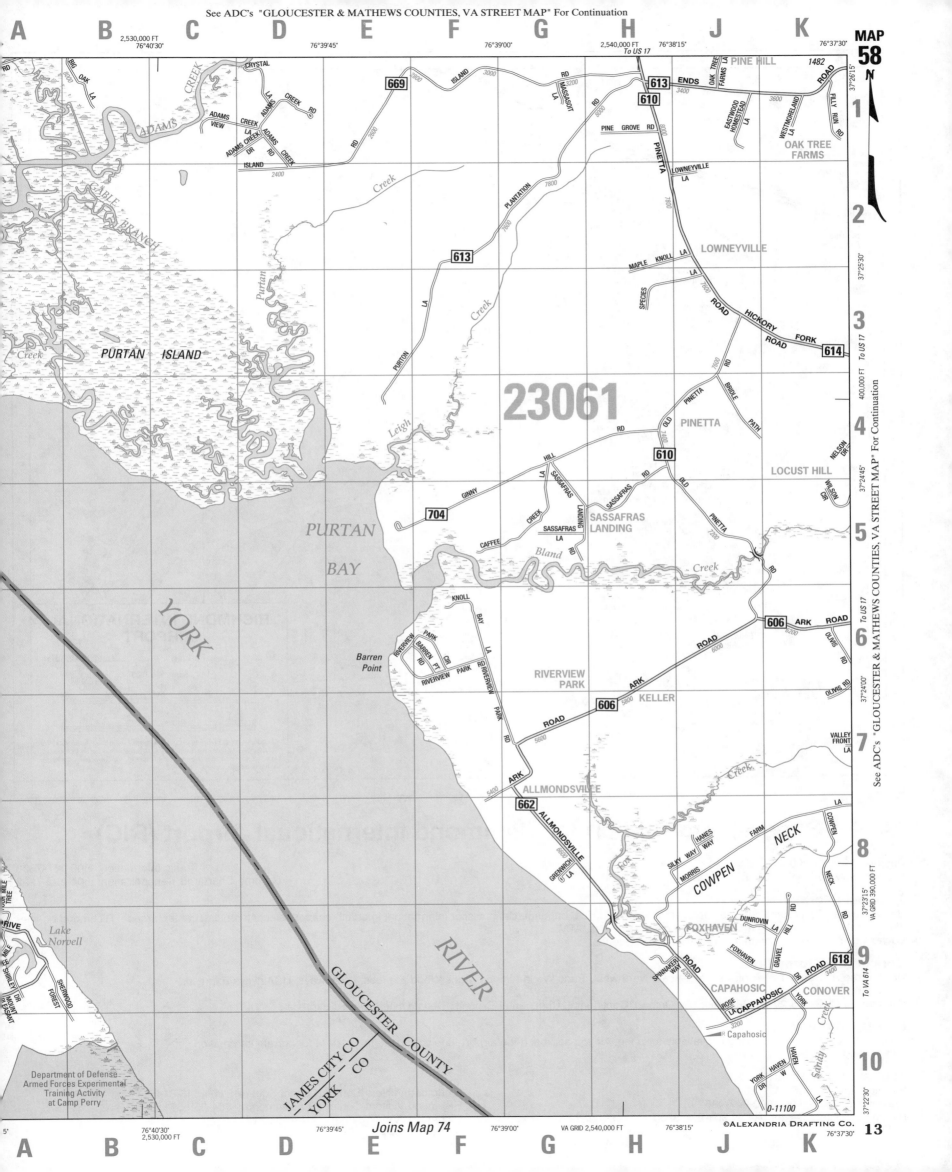

MAP 58

23061

PURTAN ISLAND

PURTAN BAY

YORK

RIVER

GLOUCESTER COUNTY

JAMES CITY CO
YORK CO

Department of Defense
Armed Forces Experimental
Training Activity
at Camp Perry

Lake Norvell

PINE HILL
OAK TREE FARMS

LOWNEYVILLE

PINETTA

HICKORY FORK ROAD

LOCUST HILL

SASSAFRAS LANDING

RIVERVIEW PARK

Barren Point

ALLMONDSVILLE

COWPEN NECK

FOXHAVEN

CAPAHOSIC

CONOVER

©Alexandria Drafting Co.

13

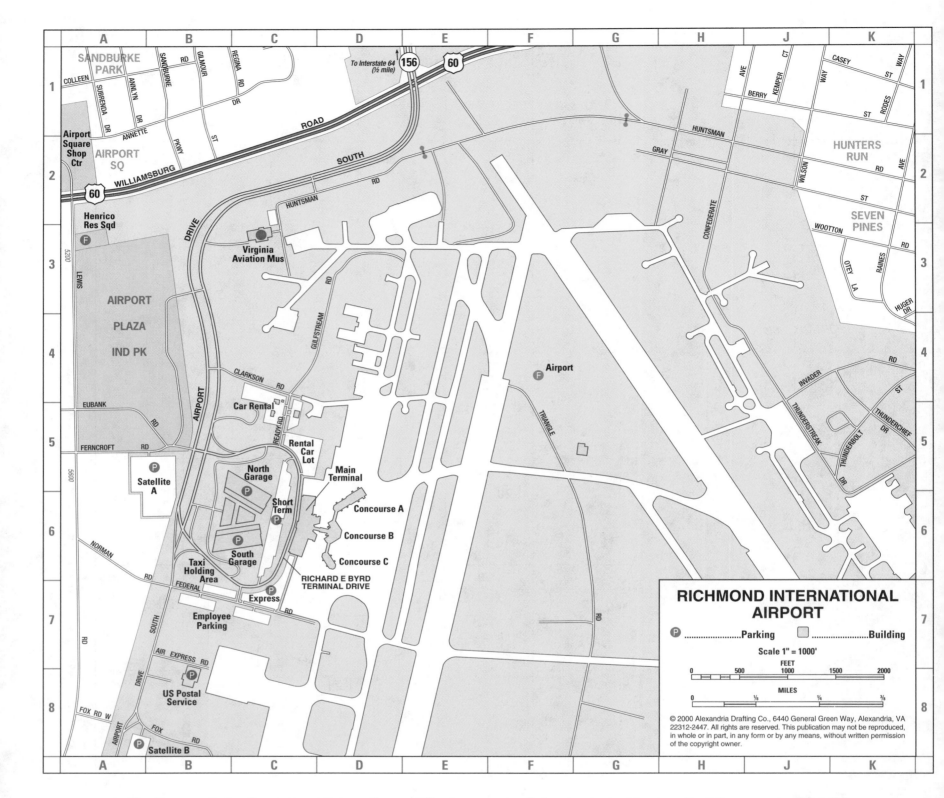

General Information for Richmond International Airport (RIC)

Phone: 804-226-3000
Internet: www.flyrichmond.com

Traffic Information: 800-367-7623
Groome Chartered Transportation: 804-222-7222

Terminal Information:
Upper Level: Rest rooms, smarte carte rental, first aid/police, baggage claim, visitor information, interfaith chapel, teleconferencing center, arcade, Richmond on-line lounge,restaurants, vending area, gift shop, pay telephones, ATM.
Lower Level: Ticket counters, car rentals, pay telephones.

Concourse Information:
Concourse A (Gates 1-8) Airlines: American, Northwest, Trans World Airlines, Trans World Express, USAirways, USAirways Express.
Concourse B Airlines: Comair.
Concourse C (Gates 20-26) Airlines: Air Ontario, Continental, Continental Express, Delta, United, United Express.

Parking Information:
Short term and long term parking is available for a fee. As you approach the airport tune your radio to 1410AM for current information and parking updates.

Tourist Information:
State of Virginia: 800-847-4882
State of North Carolina: 800-847-4862
Richmond Visitors Center (RIC): 804-236-3260

Richmond Visitors Center (Capitol Square): 804-648-3146
Richmond Visitors Center (The Diamond): 804-358-5511
Richmond Area Weather: 804-268-1212

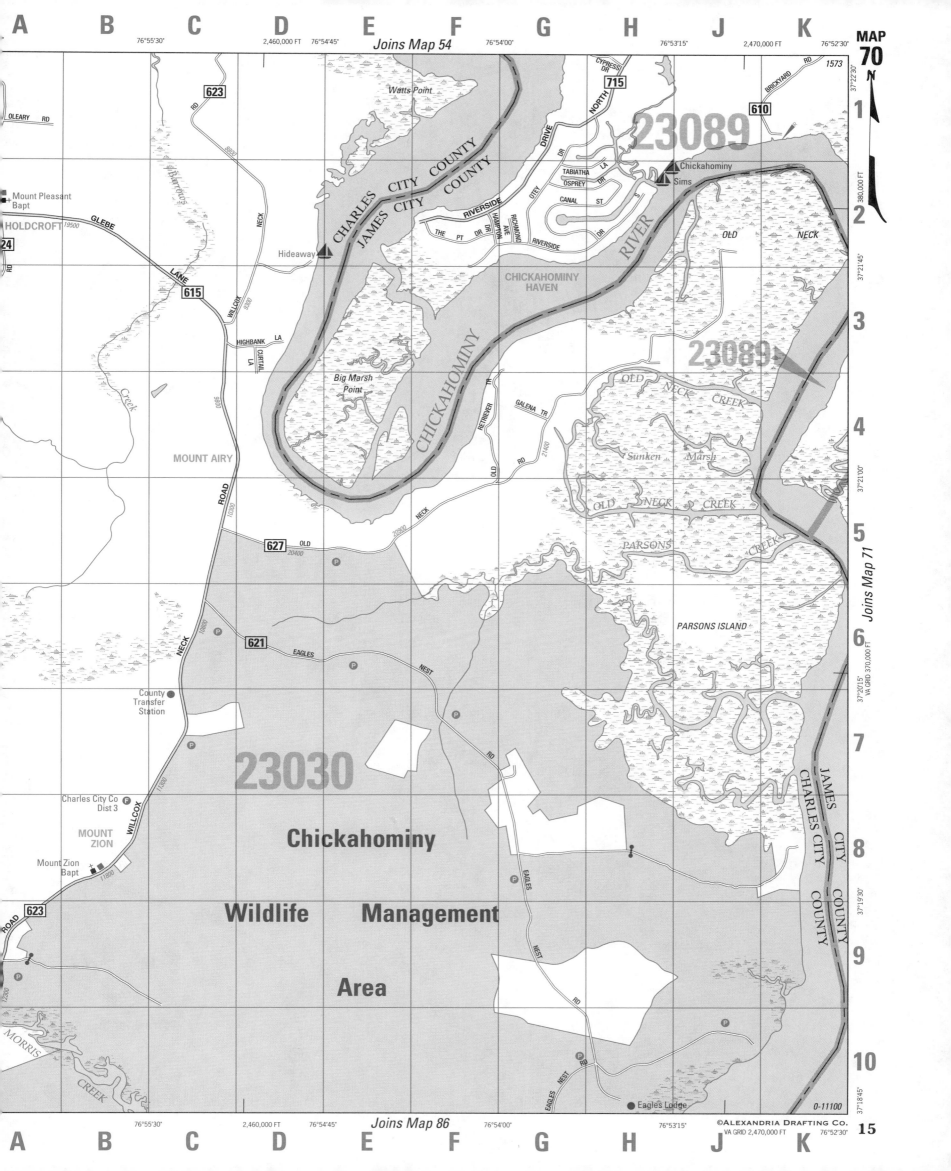

MAP
70

Joins Map 54

Joins Map 86

CHARLES CITY COUNTY
JAMES CITY COUNTY

23089

23089

23030

Chickahominy

Wildlife   Management

Area

CHICKAHOMINY RIVER

CHICKAHOMINY HAVEN

Watts Point

Hideaway

Big Marsh Point

MOUNT AIRY

OLD   NECK

OLD   NECK   CREEK

Sunken   Marsh

OLD   NECK   CREEK

PARSONS

CREEK

PARSONS ISLAND

Chickahominy
Sims

Mount Pleasant Bapt

HOLDCROFT 19500

GLEBE

OLEARY   RD

Barones   Creek

Creek

MORRIS

CREEK

Mount Zion Bapt

MOUNT ZION

County Transfer Station

Charles City Co Dist 3

Eagles Lodge

JAMES CITY COUNTY

CHARLES CITY COUNTY

CYPRESS DR

NORTH

DRIVE

OTEY DR

TABIATHA LA

OSPREY DR

CANAL ST

RICHMOND AVE

HAMPTON

THE PT DR

RIVERSIDE DR

RIVERSIDE

NECK

WILLCOX

LANE

HIGHBANK LA

CURTAIL LA

NECK ROAD

NECK

WILLCOX

RETRIEVER TR

OLD   RD

GALENA TR

EAGLES

NEST

RD

EAGLES

NEST

RD

EAGLES NEST RD

623

615

627

621

623

623

610

715

624

1573

Joins Map 71

©Alexandria Drafting Co.
VA GRID 2,470,000 FT

0-11100

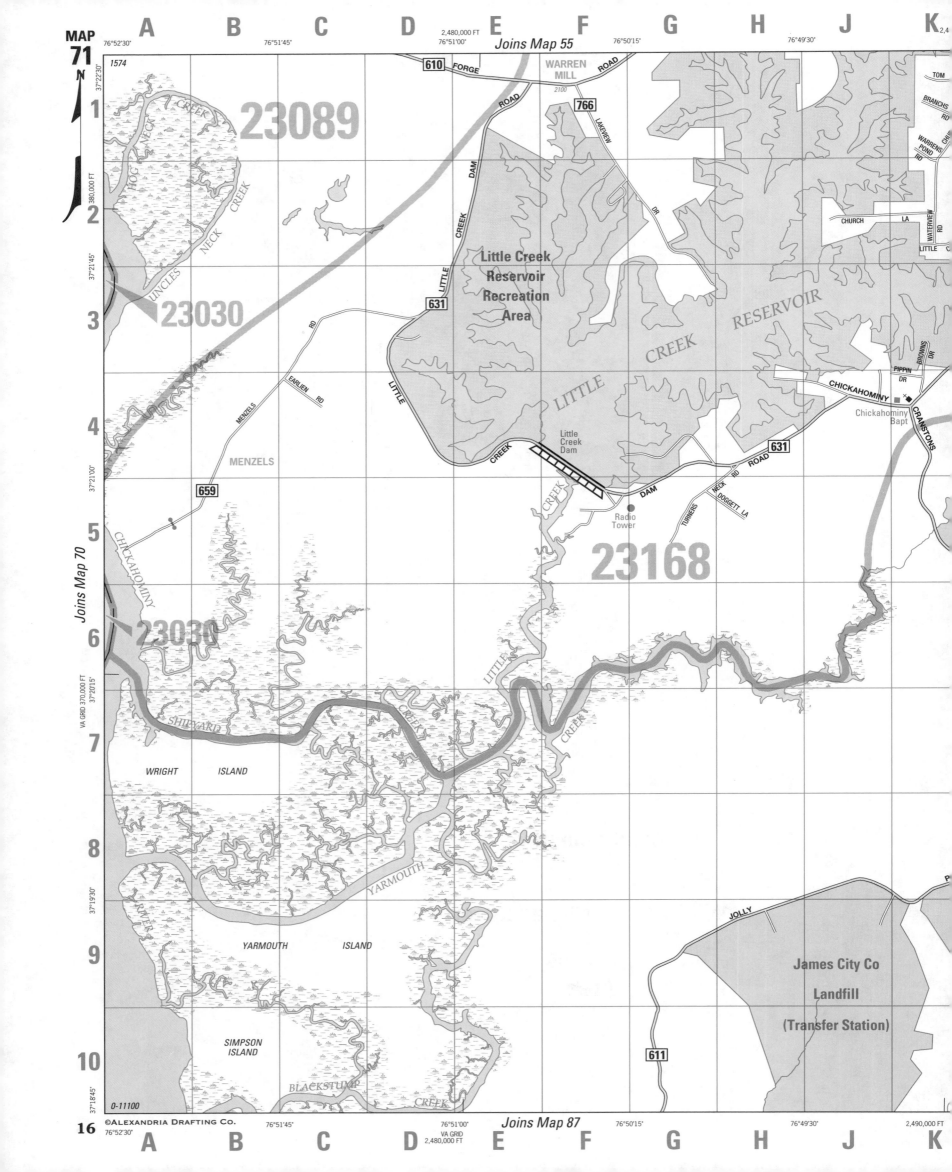

MAP
71

A B C D E F G H J K

2,480,000 FT

Joins Map 55

76°52'30" 76°51'45" 76°51'00" 76°50'15" 76°49'30"

23089

1574

610 FORGE

WARREN MILL

ROAD

2100

766

TOM

BRANCHS RD

WARRENS POND RD

LAKEVIEW DR

CHURCH LA

WATERVIEW RD

LITTLE

23030

HOG NECK

UNCLES NECK CREEK

CREEK

631

LITTLE CREEK

Little Creek
Reservoir
Recreation
Area

RESERVOIR

LITTLE

CREEK

RESERVOIR

PIPPIN DR

BROWNS DR

CHICKAHOMINY

CHICKAHOMINY BAPT

CRANSTONS

RD

Chickahominy
Bapt

EARLIEN RD

MENZELS RD

MENZELS

Little
Creek
Dam

631

659

LITTLE CREEK DAM

TURNERS NECK RD

DOGGETT LA

Radio
Tower

23168

CHICKAHOMINY

Joins Map 70

CREEK

23030

SHIPYARD

LITTLE

CREEK

WRIGHT ISLAND

CREEK

YARMOUTH

RIVER

YARMOUTH ISLAND

JOLLY

James City Co

Landfill

(Transfer Station)

SIMPSON
ISLAND

611

BLACKSTUMP

CREEK

VA GRID 370,000 FT

380,000 FT

VA GRID 2,480,000 FT

2,490,000 FT

37°22'30" 37°21'45" 37°21'00" 37°20'15" 37°19'30" 37°18'45"

0-11100

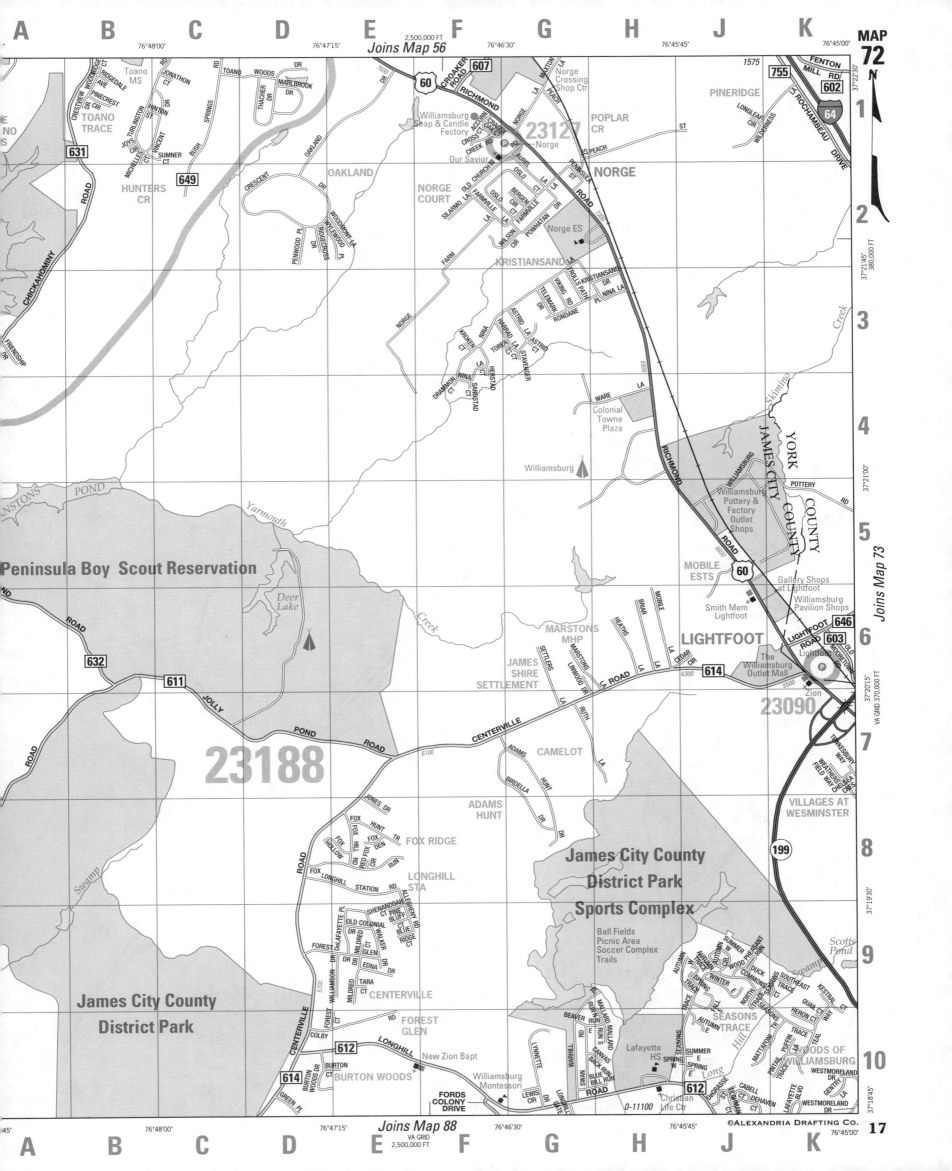

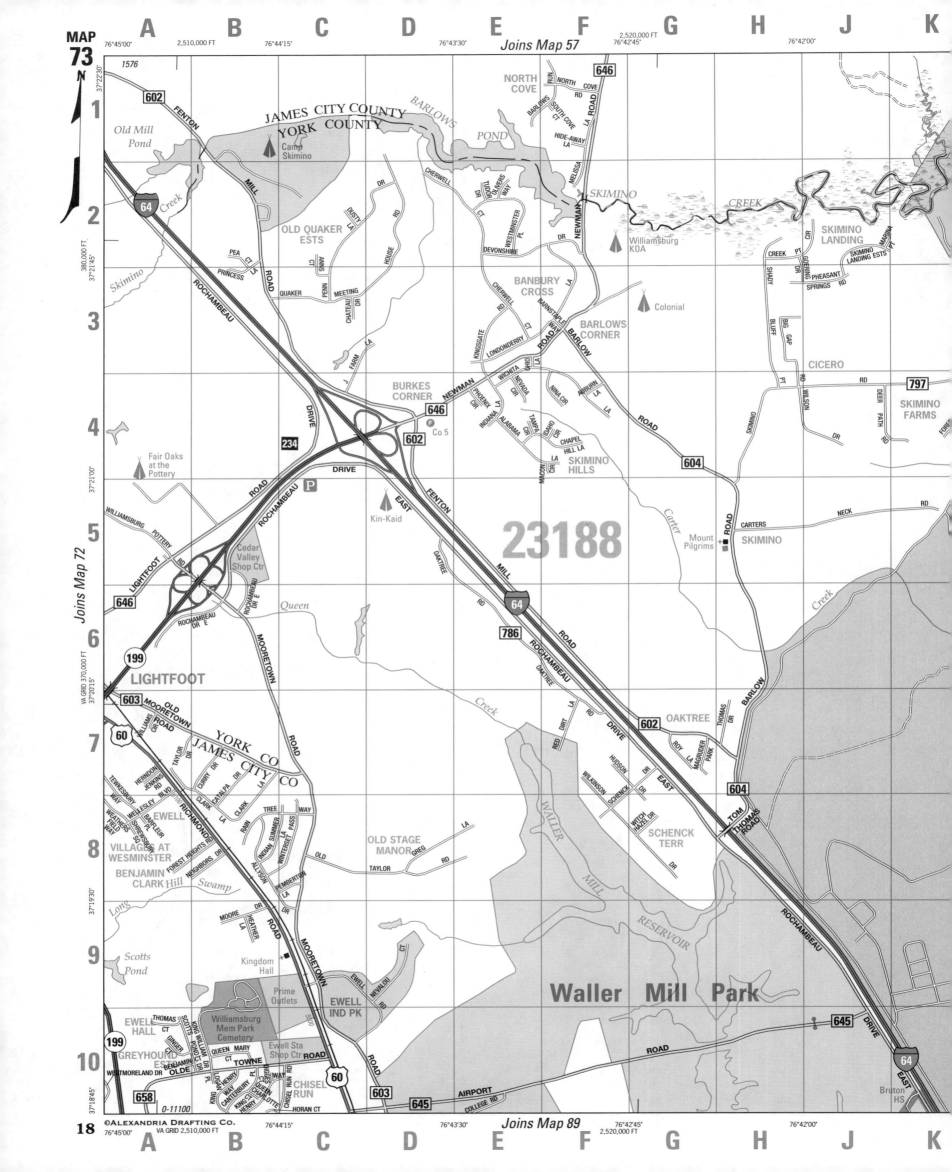

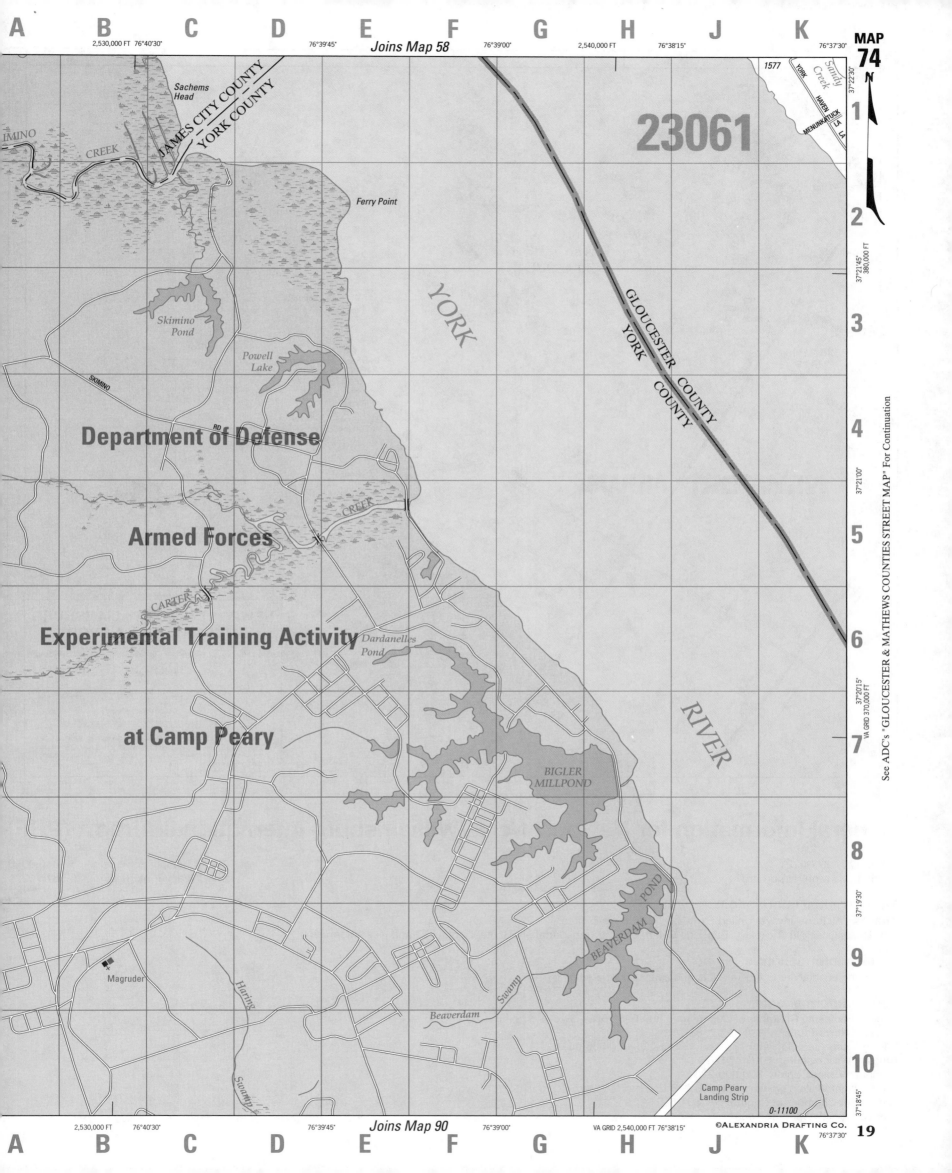

2,530,000 FT   76°40'30"   76°39'45"   *Joins Map 58*   76°39'00"   2,540,000 FT   76°38'15"

37°22'30"

1577

Sandy Creek

YORK HAVEN
MENUNKATUCK
15 LA LA

23061

Sachems Head

JAMES CITY COUNTY
YORK COUNTY

IMINO

CREEK

Ferry Point

1

2

37°21'45"
380,000 FT

YORK

GLOUCESTER COUNTY
YORK COUNTY

3

Skimino Pond

Powell Lake

SKIMINO RD

Department of Defense

4

37°21'00"

Armed Forces

CREEK

5

CARTER

Experimental Training Activity

Dardanelles Pond

6

37°20'15"
370,000 FT
VA GRID

RIVER

at Camp Peary

7

BIGLER MILLPOND

8

37°19'30"

Magruder

Haring Swamp

BEAVERDAM POND

9

Swamp

Beaverdam

10

37°18'45"

Camp Peary Landing Strip

0-11100

See ADC's "GLOUCESTER & MATHEWS COUNTIES STREET MAP" For Continuation

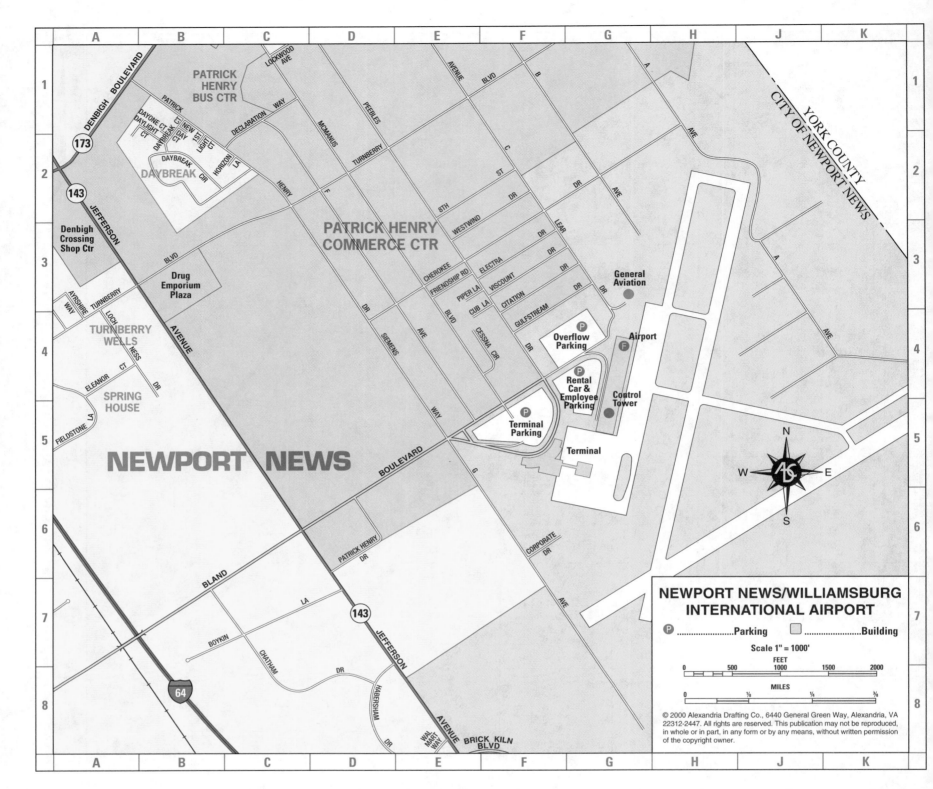

General Information for Newport News/Williamsburg International Airport (PHF)

Phone: 757-877-0221
Internet: www.phf-airport.org

Traffic Tunnel Information: 757-640-0055
Local Weather Information: 757-877-1221

Main Terminal Information:

Airlines: AirTran Airways, United Express, US Airways Express.
Services: Check-In, Ticketing, Baggage Claim, Ground Transportation, Rental Cars, USO and local information.

Parking Information:

All parking is adjacent to the Terminal and available for a small fee.

Tourist Information:

Virginia Peninsula Chamber of Commerce: 800-556-1822
Newport News Tourism: 888-493-7386
Williamsburg Area Convention & Visitors Bureau: 800-368-6511
Colonial Williamsburg: 800-447-8679
Hampton Visitor Center: 800-800-2202
Isle of Wight Visitors Center: 800-365-9339
Gloucester Chamber of Commerce: 804-693-2425
State of Virginia: 800-847-4882

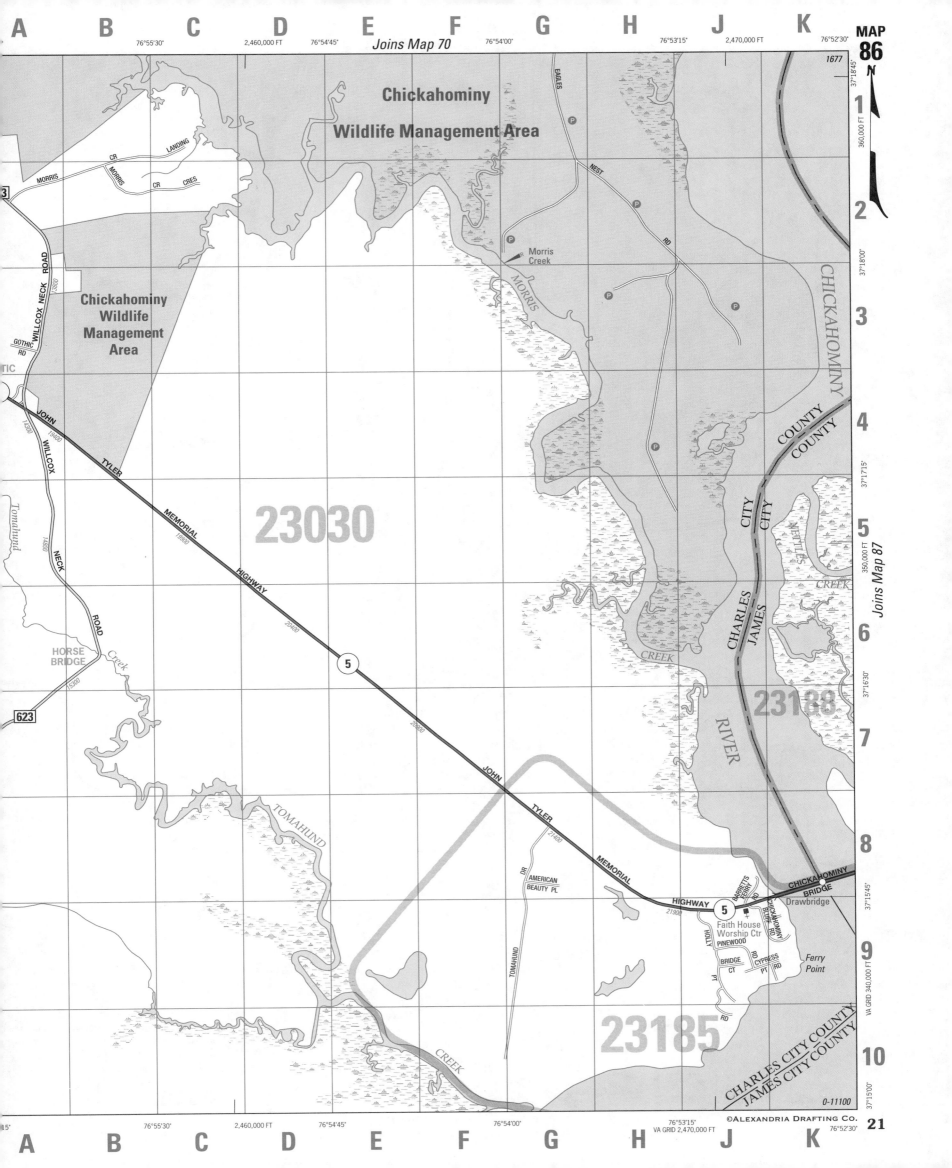

MAP 87 N

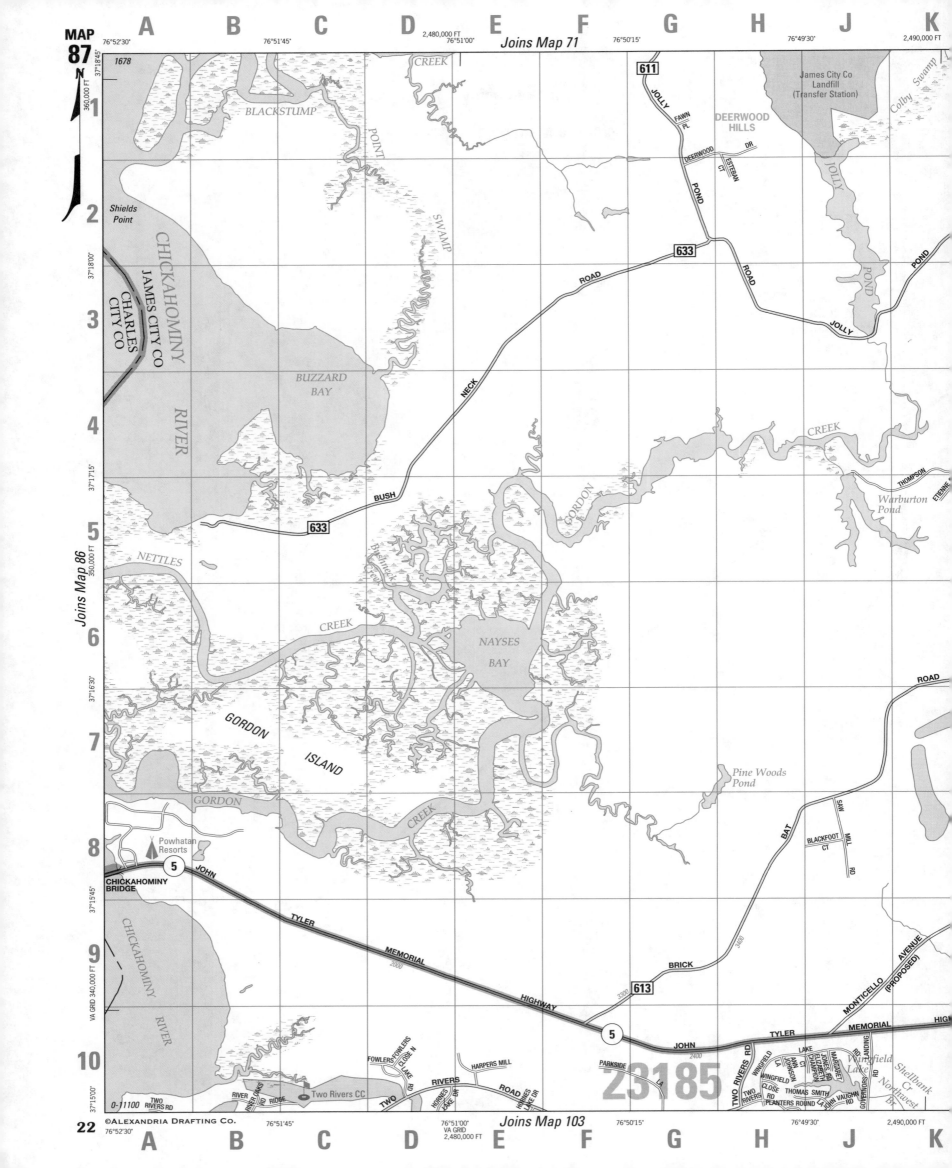

A B C D E F G H J K

76°52'30" 76°51'45" 2,480,000 FT 76°51'00" 76°50'15" 76°49'30" 2,490,000 FT

1

1678

37°18'45"

360,000 FT

BLACKSTUMP

611

JOLLY

FAWN PL

James City Co Landfill (Transfer Station)

DEERWOOD HILLS

Colby Swamp

2

Shields Point

CHICKAHOMINY

JAMES CITY CO

POINT

SWAMP

CREEK

DEERWOOD

CT

ESTEBAN

DR

633

POND

JOLLY

POND

POND

37°18'00"

3

CHARLES CITY CO

CHARLES CITY CO

RIVER

BUZZARD BAY

ROAD

NECK

ROAD

JOLLY

37°17'15"

4

RIVER

BUSH

GORDON

CREEK

THOMPSON

ETIENNE

Warburton Pond

5

633

NETTLES

Bushneck Creek

GORDON

37°16'30"

350,000 FT

6

CREEK

NAYSES BAY

ROAD

7

GORDON

ISLAND

CREEK

Pine Woods Pond

ROAD

BAT

SAW MILL RD

8

Powhatan Resorts

5

JOHN

GORDON

CREEK

BLACKFOOT CT

37°15'45"

CHICKAHOMINY BRIDGE

9

TYLER

MEMORIAL

2000

BRICK

3400

MONTICELLO (PROPOSED)

AVENUE

VA GRID 340,000 FT

37°15'15"

CHICKAHOMINY

HIGHWAY

3300

613

5

JOHN

2400

TYLER

MEMORIAL

HIGH

10

RIVER

FOWLERS

FOWLERS CLOSE N

LAKE RD

HARPERS MILL

ROAD

PARKSIDE

23185

TWO RIVERS RD

WINGFIELD

LAKE CT

JONES CT

ELIZABETH

MARGARET

CHAMPION

Wingfield Lake

RD

LANDING

Shellbank Cr

Northwest Br

37°15'00"

0-11100

TWO RIVERS RD

RIVER OAKS RD

RIDGE

Two Rivers CC

TWO RIVERS

HORNES LAKE DR

ROAD

HORNES LAKE DR

ANN JOHNSON LA

WINGFIELD CLOSE

THOMAS

SMITH LA

JOHN VAUGHN

GOVERNORS RD

©ALEXANDRIA DRAFTING CO.

76°52'30"

76°51'45"

76°51'00" VA GRID 2,480,000 FT

76°50'15"

76°49'30"

2,490,000 FT

A B C D E F G H J K

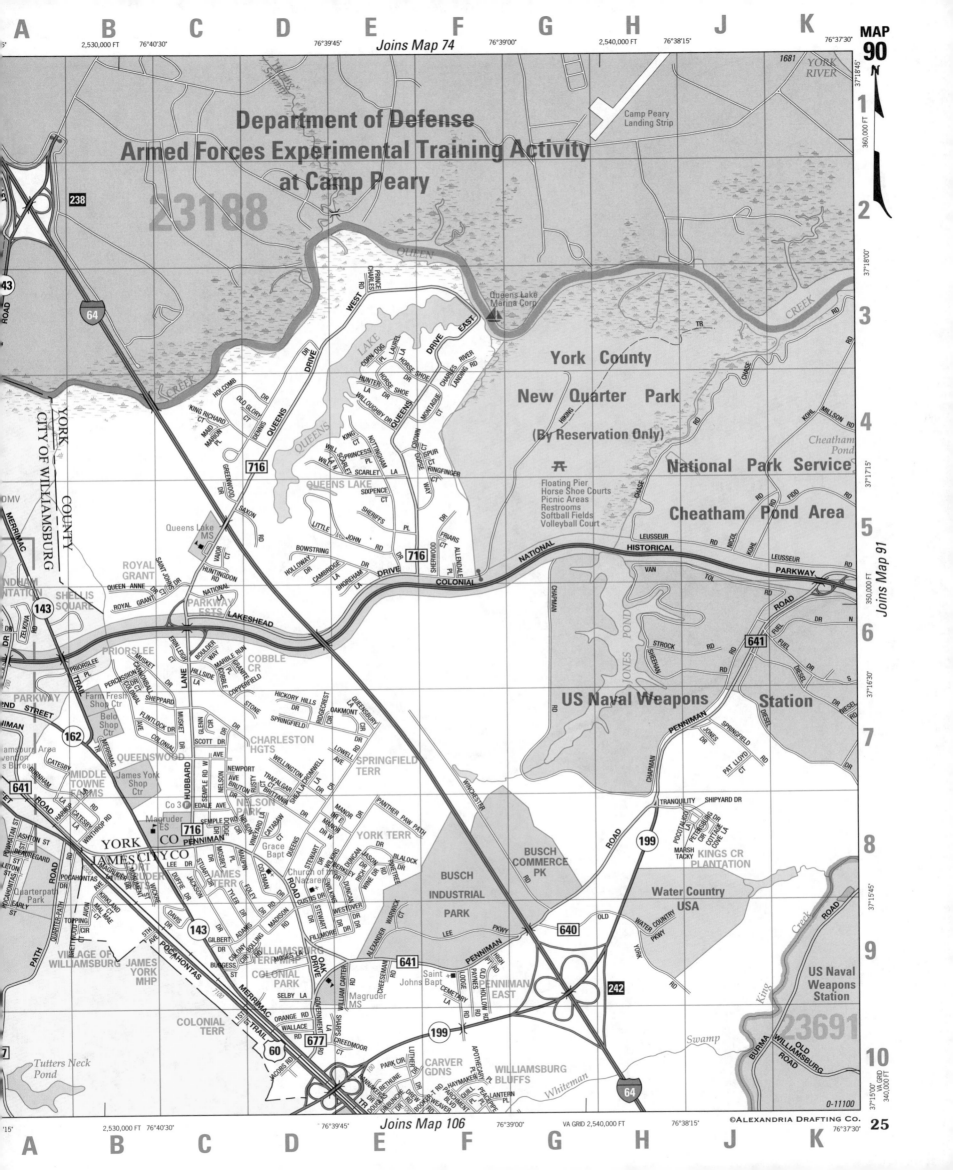

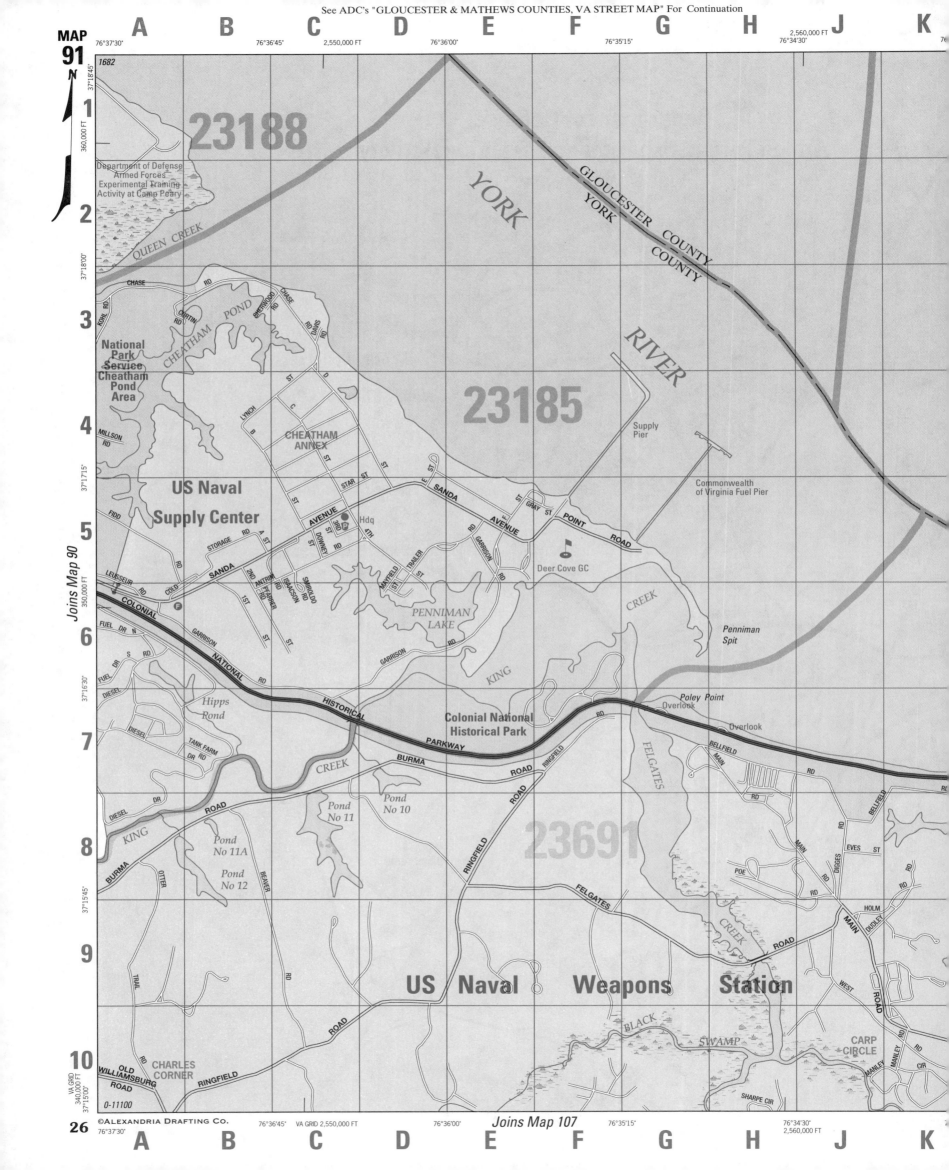

MAP
91

A B C D E F G H J K

76°37'30"   76°36'45"   2,550,000 FT   76°36'00"   76°35'15"   2,560,000 FT   76°34'30"

1

1682
37°18'45"

360,000 FT

Department of Defense
Armed Forces
Experimental Training
Activity at Camp Peary

2

37°18'00"

QUEEN CREEK

YORK

GLOUCESTER COUNTY

YORK COUNTY

CHASE

RD

KOHL RD

CHEATHAM POND

CURTIN RD

SHERWOOD RD

CHASE RD

DAVIS RD

3

37°17'30"

National
Park
Service
Cheatham
Pond
Area

CHEATHAM POND

ST

C

D

LYNCH

RIVER

23185

RD

4

MILLSON RD

37°17'15"

B

CHEATHAM
ANNEX

ST

ST

E ST

ST

STAR ST

SANDA

Supply
Pier

Commonwealth
of Virginia Fuel Pier

US Naval

Supply Center

FIDD

AVENUE

3RD

ST

Hdq

4TH

GRAY ST

ST

AVENUE

POINT

ROAD

5

350,000 FT

LEUSSEUR RD

STORAGE RD

A ST

DOWNEY

ST

RD

RD

GARRISON RD

RD

Deer Cove GC

Joins Map 90

SANDA

2ND

ANTRUM

1ST

ISAACSON

SMIROLDO RD

MAYFIELD ST

TRAILER ST

GARRISON

RD

CREEK

Penniman
Spit

COLD

PFAHLER

RD

COLONIAL

ST

RD

PENNIMAN
LAKE

RD

6

37°16'30"

FUEL DR N

GARRISON

NATIONAL

ST

KING

FUEL DR S

RD

RD

GARRISON

RD

Poley Point
Overlook

DIESEL

HISTORICAL

Colonial National
Historical Park

RD

Overlook

BELLFIELD

MAIN

RD

7

37°16'15"

Hipps
Pond

PARKWAY

BURMA

ROAD

RINGFIELD

FELGATES

RD

DIESEL

CREEK

TANK FARM

DR RD

Pond
No 10

ROAD

MAIN

RD

BELLFIELD

8

DIESEL DR

ROAD

Pond
No 11

23691

RD

RD

EVES ST

37°15'45"

KING

BURMA

Pond
No 11A

BEAVER

RINGFIELD

FELGATES

POE RD

DIGGES RD

MAIN RD

RD

OTTER

Pond
No 12

ROAD

9

TRAIL

RD

US Naval Weapons Station

FELGATES ROAD

WEST ROAD

MAIN ROAD

HOLM

DUDLEY

10

37°15'00"

OLD
WILLIAMSBURG
ROAD

CHARLES
CORNER

RINGFIELD

ROAD

BLACK

SWAMP

CARP
CIRCLE

MANLEY RD

CIR

SHARPE CIR

VA GRID
340,000 FT

0-11100

76°37'30"   76°36'45"   VA GRID 2,550,000 FT   76°36'00"   Joins Map 107   76°35'15"   76°34'30"   2,560,000 FT

A B C D E F G H J K

MAP 92

23072

23184

WICOMICO

23062

27

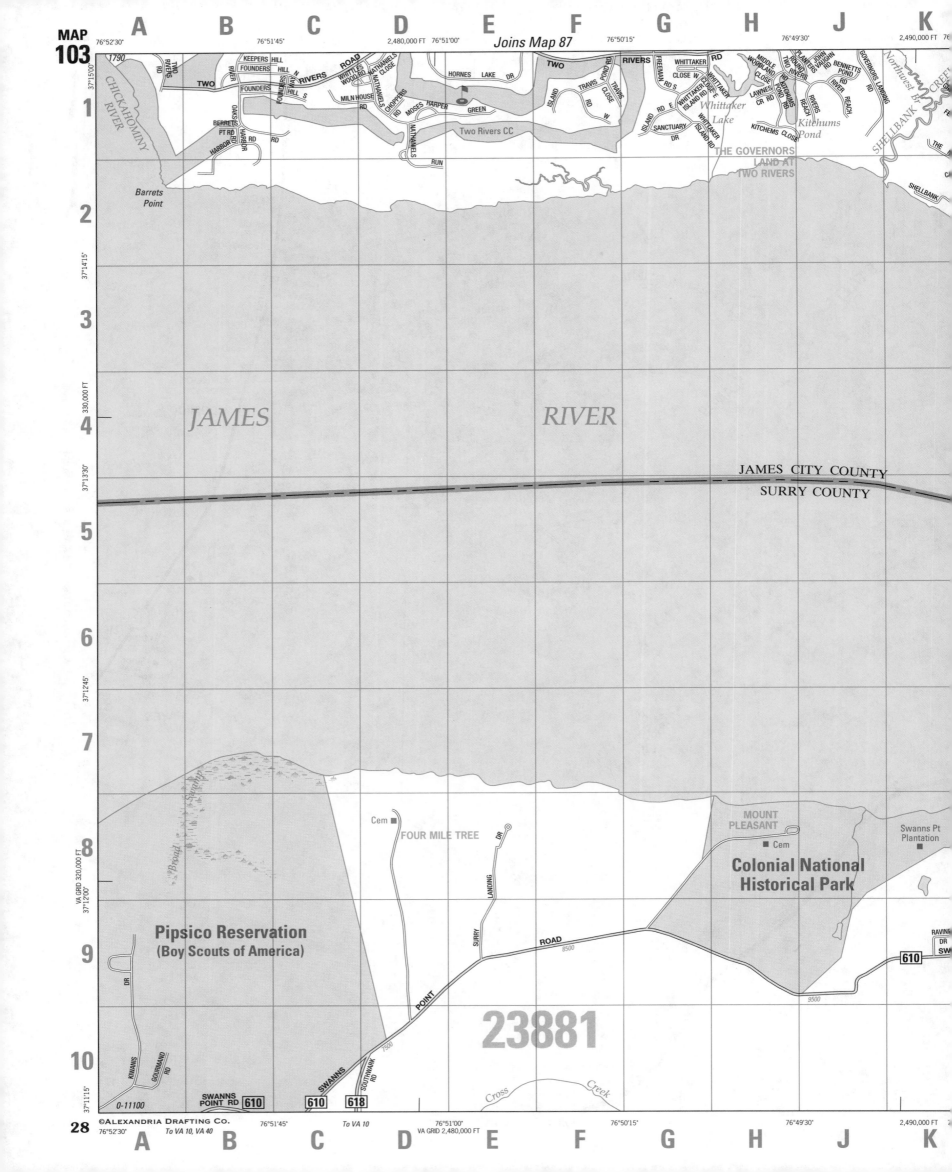

MAP
103

Joins Map 87

A B C D E F G H J K

76°52'30" 76°51'45" 2,480,000 FT 76°51'00" 76°50'15" 76°49'30" 2,490,000 FT 76°

1790

CHICKAHOMINY RIVER

TWO RIVERS RD

KEEPERS HILL
FOUNDERS

TWO RIVERS

FOUNDERS HILL

RIVERS ROAD

WHITTLES WOOD RD
NATHANIELS CLOSE

HORNES LAKE DR

TWO RIVERS

WHITTAKER RD

FREEMAN CLOSE W
WHITTAKER CLOSE W

MIDDLE WOODLAND CLOSE
PLANTERS RD
TWO RIVERS

JOHN VAUGHN RD

GOVERNORS LANDING RD

Northwest Br

BERRETS PT RD
HARBOR RD
RD

FOUNDERS HILL S

MILN HOUSE

CHOPPERS RD

MOSES HARPER

GREEN

ISLAND RD E
ISLAND RD
SANCTUARY DR

WHITTAKER ISLAND RD E
WHITTAKER ISLAND RD

KITCHUMS CLOSE RD
KITCHUMS POND RD

RIVERS REACH
RIVER REACH

SHELLBANK CREEK

NATHANIELS RD

RUN

Two Rivers CC

Whittaker Lake

Kitchums Pond

THE GOVERNORS LAND AT TWO RIVERS

SHELLBANK

Barrets Point

JAMES   RIVER

JAMES CITY COUNTY
SURRY COUNTY

Broad Swamp

Cem

FOUR MILE TREE

LANDING DR

SURRY LANDING

MOUNT PLEASANT
Cem

Colonial National Historical Park

Swanns Pt Plantation

RAVINE DR
SW

Pipsico Reservation
(Boy Scouts of America)

DR

ROAD
8500

610

SURRY POINT

9500

KIWANIS
GOURMAND RD
DR

0-11100

SWANNS POINT RD
610

SWANNS

610

618

SOUTHWARK RD

750

23881

Cross   Creek

To VA 10, VA 40

To VA 10

76°52'30"

76°51'45" 76°51'00"
VA GRID 2,480,000 FT

76°50'15" 76°49'30" 2,490,000 FT

A B C D E F G H J K

37°15'00" 37°14'15" 330,000 FT 37°13'30" 37°12'45" VA GRID 320,000 FT 37°12'00" 37°11'15"

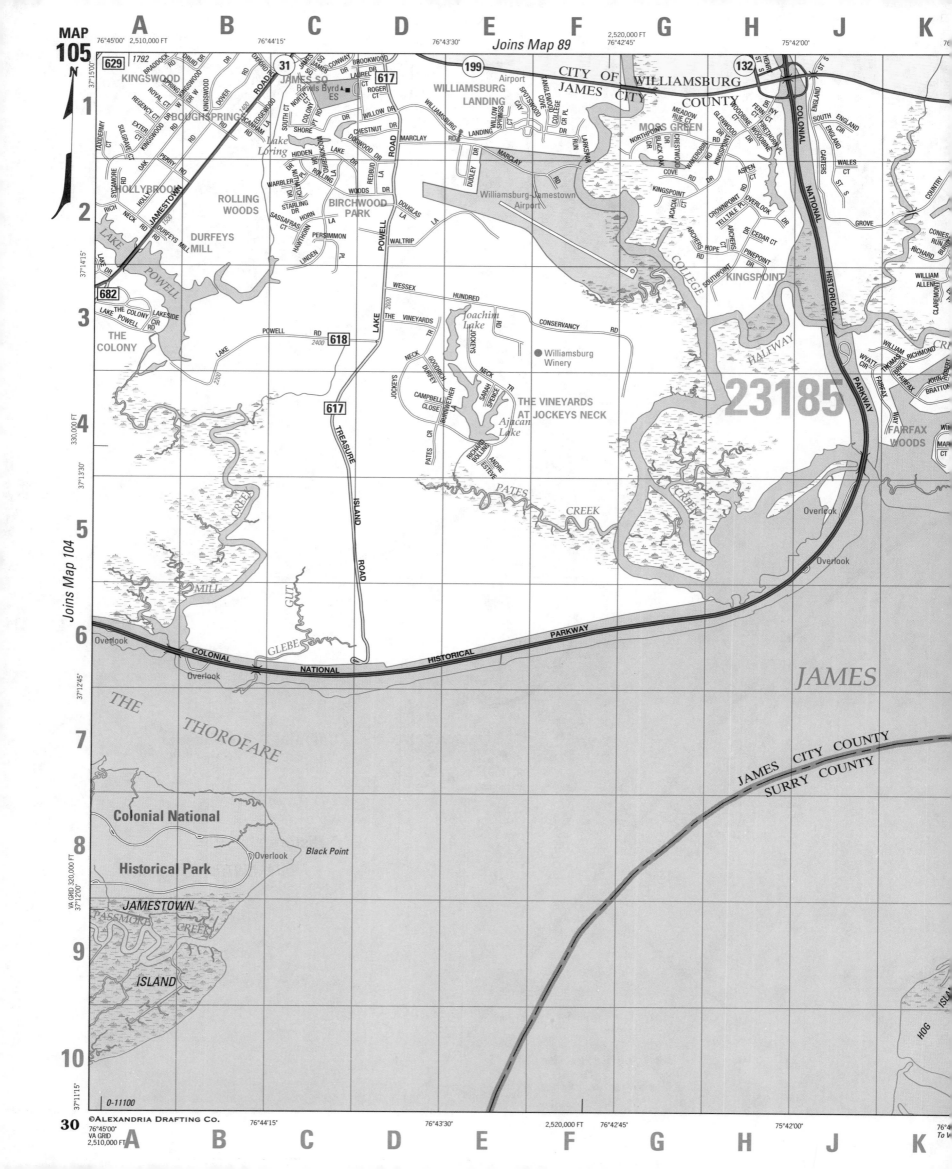

MAP 105

N

629 1792
KINGSWOOD
BRADDOCK
DRUID DR
OXFORD RD

31

JAMES SQ
CONWAY
BROOKWOOD
199
Airport
CITY OF WILLIAMSBURG
132

SPRING
KINGSWOOD
DOVER
SOUTH CT
COLONY
NORTH
LAUREL
617
ROGER
CT
JAMES CITY COUNTY
HENRI
MOSS GREEN
WALES

REGENCY CT
EXTER
KINGSWOOD
RD
PERRY
RD
CANGHAM
SHORE
DR
CHESTNUT
WILLOW DR
ROGER
CT
WILLIAMSBURG
LANDING
WILLOW
SPRINGS
COLLEGE
CR PL
MEADOW
RUE CT
NORTHPOINT
DR
GLENWOOD
KINGSWOOD

TAXIDERMY
CT
SILGRAVE
CT
SYCAMORE
OAK
ROYAL
CT
Lake
Loring
HIDDEN
DR
MARCLAY
RD
LANDING
DUDLEY
DR
MARCLAY
SPOTSWOOD
ANGLEWOOD
COVE
LARKSPAR
RUN
BLACK OAK
CHESTWOOD
DR
WOODBINE
WOODORE
ASPEN

HOLLYBROOK
RICH
NECK
RD
HOLLYBROOK
DURFEYS MILL
ROLLING
WOODS
WARBLER DR
STARLING
DR
SASSAFRAS
CT
HAWTHORN
BIRCHWOOD
PARK
MOCKINGBIRD LA
ROLLING DR
WOODS
REDBUD
DR
DOUGLAS
LA
COVE
KINGSPOINT
ACACIA
CT
CROWNPOINT
OVERLOOK
CEDAR CT
WILLIAM
ALLEN

LAKE DR
LAKE
POWELL
DURFEYS MILL RD
LINDEN
PERSIMMON
PL
POWELL
RD
WALTRIP
LA
ARCHERS
HOPE
RD
ARCHERS
PINEPOINT
TELLTALE PL
CONIES
RUN
RICHARD
CLAREMONT

682
THE COLONY
LAKESIDE
CIR
LAKE
POWELL CIR
RD
POWELL
RD
618
2400
LAKE
WESSEX
THE
VINEYARDS
TR
HUNDRED
RD
Joachim
Lake
CONSERVANCY
RD
Williamsburg
Winery
SOUTHPOINT
KINGSPOINT
HALFWAY
WILLIAM
BRICE
RICHMOND

THE
COLONY
LAKE
2200
POWELL
RD
JOCKEYS
NECK
TR
NECK
TR
GOODRICH
DURFEY
CAMPBELL
CLOSE
SARAH
SPENCE
23185
WYATT
CIR
THOMAS
FAIRFAX
JOHN
BRATTON

617
TREASURE
ISLAND
ROAD
JOCKEYS
NECK
PATES
CR
BURWMEWETHER
RICHARD
BOLLING
ANDRE
ESTEVE
Ajacan
Lake
THE VINEYARDS
AT JOCKEYS NECK
FAIRFAX
WAY
FAIRFAX
WOODS
MARI
CT

CREEK
MILL
GUT
GLEBE
PATES
CREEK
Overlook
Overlook
CRIBB
Overlook

COLONIAL
NATIONAL
HISTORICAL
PARKWAY
Overlook
Overlook
JAMES

THE
THOROFARE
JAMES CITY COUNTY
SURRY COUNTY

Colonial National
Overlook
Black Point

Historical Park

JAMESTOWN
PASSMORE
CREEK

ISLAND
HOG
ISLA

©ALEXANDRIA DRAFTING CO.

76°45'00"
VA GRID
2,510,000 FT

0-11100

76°44'15"
76°43'30"
2,520,000 FT
76°42'45"
75°42'00"
76°4
To W

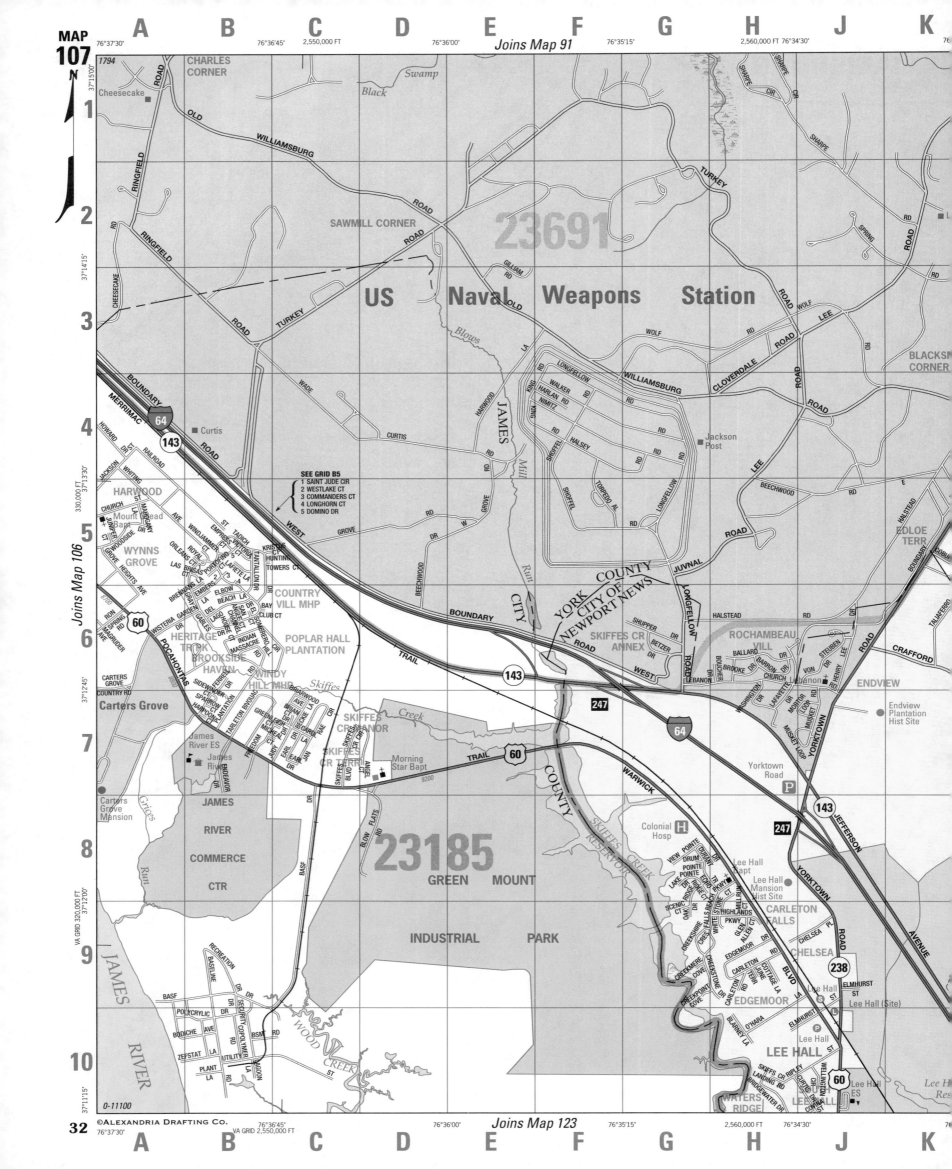

MAP
109

A B C D E F G H J K

76°30'00" 76°29'15" 2,590,000 FT 76°28'30' 76°27'45" 76°27'00" 2,600,

SPENCER RD 1796 Quarter Point Gaines Point

Virginia Inst of Marine Science (College of William & Mary)

Aquarium & Vis Ctr

23062 **23072**

YORK *RIVER*

GLOUCESTER COUNTY
YORK COUNTY

SEE YORKTOWN ENLARGEMENT

Point of Rocks

Redoubts 9 & 10

MOORE HOUSE ROAD

23690

Joins Map 108

National Cem

Moore House

238

WASHINGTON ROAD

Colonial National Historical Park

FUEL FARM RD

US Coast Guard Reserve Training Center

US Naval Supply Center

Wormley Pond

WEST BRANCH WORMLEY CREEK

Battlefield Tour (Follow Red Arrows)

NORTH BEACH RD

Wormley Cr

VA Power

631 WATERVIEW

AMERICAN OIL REFINERY

704

HISTORICAL TOUR

COOK RD
SURRENDER RD

634

MARLBANK FARM

MARLBANK

Wormley Cr Rd

WATERVIEW ROAD

WORMLEY CREEK

BACK CREEK

ROAD 173

York HS

NELSON HGTS

OLD YORK ROAD

693

632

23692

631

173

Yorktown (Little League Park)

BATTLE PARK
York Co

GEORGE WASHINGTON MEMORIAL

HAMPTON

GOFFIGAN GDNS

MARLBANK COVE

HORNSBYVILLE ROAD

Hornsbyville Meth

718

630

GOODWIN ROAD

23696

LUDLOW DR
SOMMERVILLE

Saint Joan of Arc Cath

17

HARRIS GROVE

SETTLERS CROSSING

HORNSBYVILLE

HARRIS GROVE

HORNSBYVILLE

Co 6
718 BACK CREEK

WOLF TRAP ROAD

SOMMER

EDGEHILL

BURNT BRIDGE RUN

105

PATRIOTS SQUARE

Patriot Square Shop Ctr

YORK RIVER COMMERCE PK

634

718

York Assembly of God 0-11100

York Co Operations Ctr

Seaford ES

SEAFORD ROAD

LEWIS
CLARK

76°30'00" 76°29'15" 76°28'30" VA GRID 2,590,000 FT 76°27'45" 76°27'00" 2,600,000 FT

A B C D E F G H J K

MAP
110

YORK RIVER

CHESAPEAKE

BAY

Tue Point

ISLANDS

GOODWIN

THOROFARE

Dandy Bapt

DANDY

Goodwin Neck Ests

Barcannore La

Dandy View La

Sandbox La

Belvin La

Dandy La

Ignoble La

Loop Rd

Tue Marsh La

Bradley Dr

Dandy Haven La

Middle Dandy Loop

Goodwin Neck Rd

Land Grant Rd

Stillwater

Buckingham Dr

Eisenhower Dr

BACK   CREEK

R M Mills

Berkley Beach

Ironmonger Rd

Shirley

Hidden

Back Cr

Purgold Rd

Mary Ann Dr

Melba Ct

Claxton Rd

Sunset Dr

Montgomery La

CLAXTON CREEK

Harbor Cres

Harbor Terr

Hansford Ct

718

SEAFORD

Back Creek Road

Landing

Fox Woods Rd

718

Back Cr Rd

Zion

Club Way

John Way

Wornom Dr

Seaford

Dawson Park Cres

Kenneth Dr

Blanton Dr

Raymond Dr

Cove Dr

622

Seaford Shores

SEAFORD SHORES

Old Seaford Rd

Evergreen Shores

Hansford La

712

BAY TREE BEACH ROAD

Bay Tree Point

Yorkshire Park

McPherson Ct

Dawson Dr

Seaford Ch of Christ

Goose Cr Rd

Goose Creek

August Dr

Chismans Pt

Sparrer Rd

Crockett Rd

Evergreen Rd

Smoots Rd

Charles Cir

Jan La

Wiley Rd

De Alba Rd

ROAD

York Point Road

Bay Tree Creek

Cheadle Hgts

Joanne Dr

Cheadle La

Seaford Sue Dr

1797

2,610,000 FT

340,000 FT

330,000 FT

VA GRID 320,000 FT

0-11100

76°25'30"   76°24'45"   76°24'00"   76°23'15"   76°22'30"

37°15'00"   37°14'15"   37°13'30"   37°12'45"   37°12'00"   37°11'15"

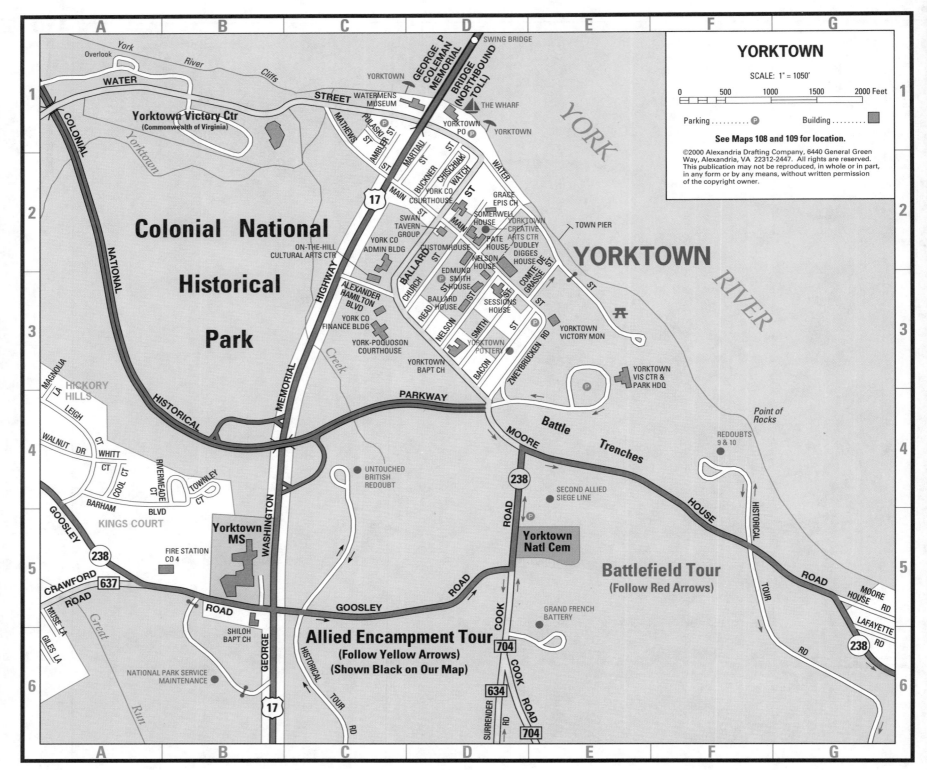

Points of Interest

Yorktown, founded in 1691, was one of only a few colonial ports authorized to ship tobacco from the colonies. Today Yorktown is better known as the site of the last major battle in the Revolutionary War and the symbolic end of British dominance over North America. General Cornwallis surrendered British forces to a combined French and American force commanded by General George Washington at Yorktown on October 19, 1781.

Ballard House — D-3
The Ballard House is the restored home of Captain John Ballard, merchant and sea captain from 1727-1744.

Customhouse — D-2
The Customhouse was reputed to have been built in 1721 as Richard Ambler's "large brick storehouse" and used as his official office while he served as collector of customs in Yorktown. It is currently owned by the Comte de Grasse chapter, Daughters of the American Revolution.

Dudley Digges House — D-2
The Dudley Digges House is the restored townhouse built by the Digges family in the early 18th century. C. Dudley Digges served as a council member for the state of Virginia during the Revolutionary Era.

Earthworks at Colonial National Historical Park — F-4
Near Yorktown are the remains of the many British earthworks of 1781. These same fortifications redoubts were modified and used by Confederate forces during the peninsula campaign of the Civil War. Because Washington ordered the dismantling of the French and American works following the siege, these have been reconstructed after careful archeological examination and discriminating research.

Edmund Smith House — D-3
The Edmund Smith House is the restored home named for the builder who

willed the house to his daughter, Mildred, wife of David Jamerson. In 1781, Lieutenant Governor Jamerson lived beside wartime General Thomas Nelson.

Grace Episcopal Church — D-2
The Grace Episcopal Church, originally known as the York-Hampton Church, was erected in 1679. Although it was damaged several times by fire, it was later rebuilt with its original walls. Today it continues to be used by the community.

Moore House — G-6
Beautifully restored, the Augustine Moore house was the site of the negotiations between the British, French and Americans. The peace commission met here on October 18 and 19 of 1781 to draft and then ratify the terms under which Cornwallis would surrender the British forces to General Washington.

Nelson House — D-2
The Nelson House is the restored mansion built by "Scotch Tom" Nelson in the early 18th century. This impressive example of Georgian architecture was the home of his grandson, Thomas Nelson, Jr., a signer of the Declaration of Independence.

Pate House — D-2
The Pate House was the original home named for the first owner, Thomas Pate. The home was built at the turn of the 18th century and later sold to the Digges family.

Sessions House — D-2
Reputed to be the oldest house in Yorktown, the Sessions House was built by Thomas Sessions in 1692. The home survived the siege of 1781.

Somerwell House — D-2
The Somerwell House is the restored brick home of Mungo Somerwell, one of

Yorktown's ferrymen. The home survived the siege of 1781 and the Civil War. The house was restored in 1936.

Swan Tavern Group — C-2
This reconstructed tavern and dependencies are built on their original sites. The popular tavern was built in 1722 by "Scotch Tom" Nelson and Joseph Walker. Having survived the siege of 1781, it was later wrecked during the Civil War.

Watermen's Museum — C-1
Located on the Yorktown Waterfront, the Watermen's Museum serves to preserve and interpret the rich heritage of the Watermen of the Chesapeake Bay.

Yorktown Victory Center — A-1
Through exhibits featuring 18th century artifacts and documentary film, the story of the American Revolution is brought to life for the visitor. Life in a typical Continental Army encampment is presented by costumed docents demonstrating colonial medicine, cooking and military techniques.

Yorktown Victory Monument — E-3
The Yorktown Victory Monument was erected in commemoration of the French Alliance and the victory over Cornwallis. The cornerstone of this statue was laid in 1881 during the celebration of the centennial of the surrender.

Yorktown Visitor Center — E-3
Run by the National Park Service, the Yorktown Visitor Center offers a series of dioramas depicting the siege; a theater program tells the story of the town, and the center has many exhibits, including military tents used by General Washington.

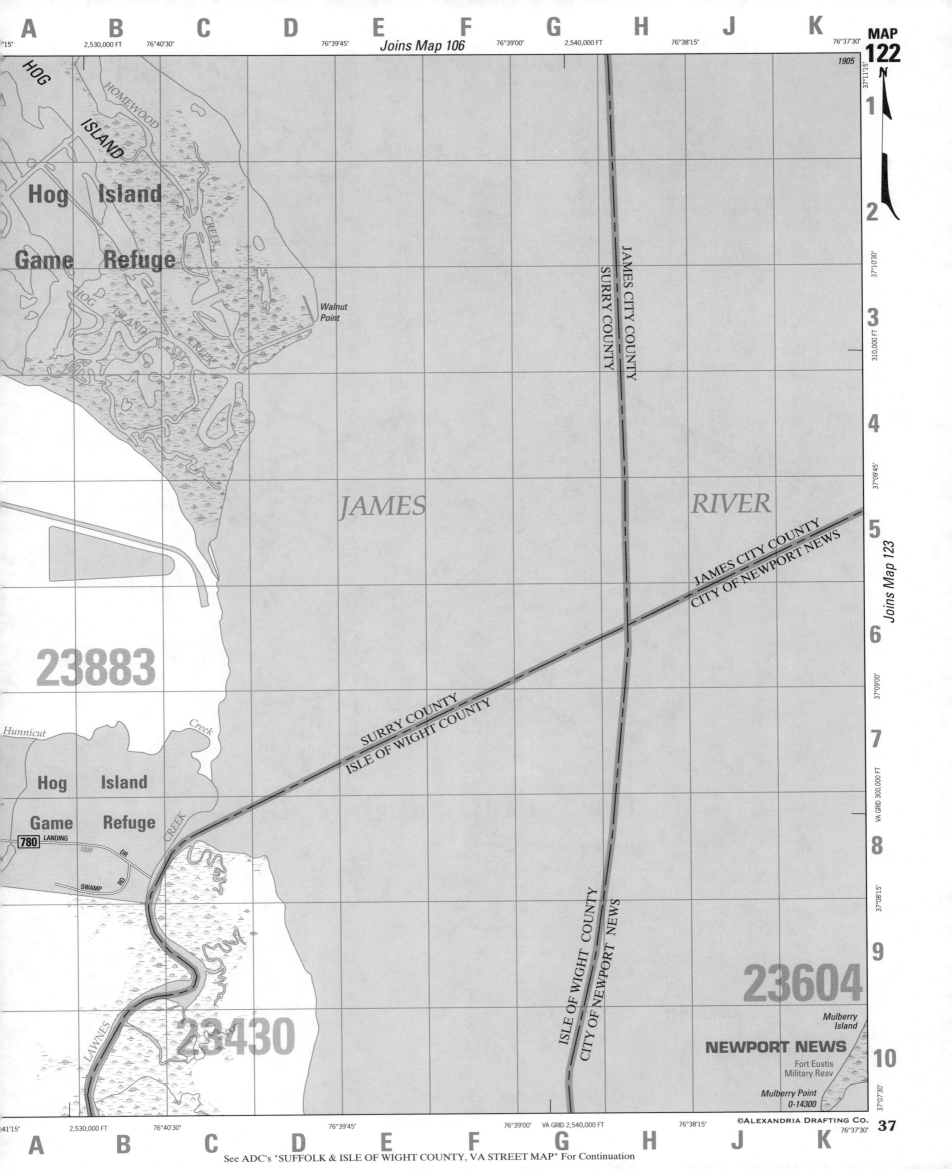

MAP
122
1905

A   B   C   D   E   F   G   H   J   K

"15'   2,530,000 FT   76°40'30"   76°39'45"   76°39'00"   2,540,000 FT   76°38'15"   76°37'30"

37°11'15"

HOG

HOMEWOOD

ISLAND

Hog   Island

Game   Refuge

HOG   ISLAND CREEK

CREEK

Walnut Point

37°10'30"

37°10'

JAMES   *RIVER*

JAMES CITY COUNTY
SURRY COUNTY

37°09'45"

JAMES CITY COUNTY
CITY OF NEWPORT NEWS

23883

310,000 FT

Hunnicut   Creek

SURRY COUNTY
ISLE OF WIGHT COUNTY

37°09'00"

Hog   Island

Game   Refuge

780   LANDING

1000   DR

RD

SWAMP

CREEK

VA GRID 300,000 FT

37°08'15"

ISLE OF WIGHT COUNTY
CITY OF NEWPORT NEWS

LAWNES

23430

23604

Mulberry Island

NEWPORT NEWS

Fort Eustis
Military Resv

Mulberry Point
0-14300

37°07'30"

ISLE OF WIGHT COUNTY
CITY OF NEWPORT NEWS

©Alexandria Drafting Co.

A   B   C   D   E   F   G   H   J   K

41'15"   2,530,000 FT   76°40'30"   76°39'45"   76°39'00"   VA GRID 2,540,000 FT   76°38'15"   76°37'30"

1   2   3   4   5   6   7   8   9   10

37

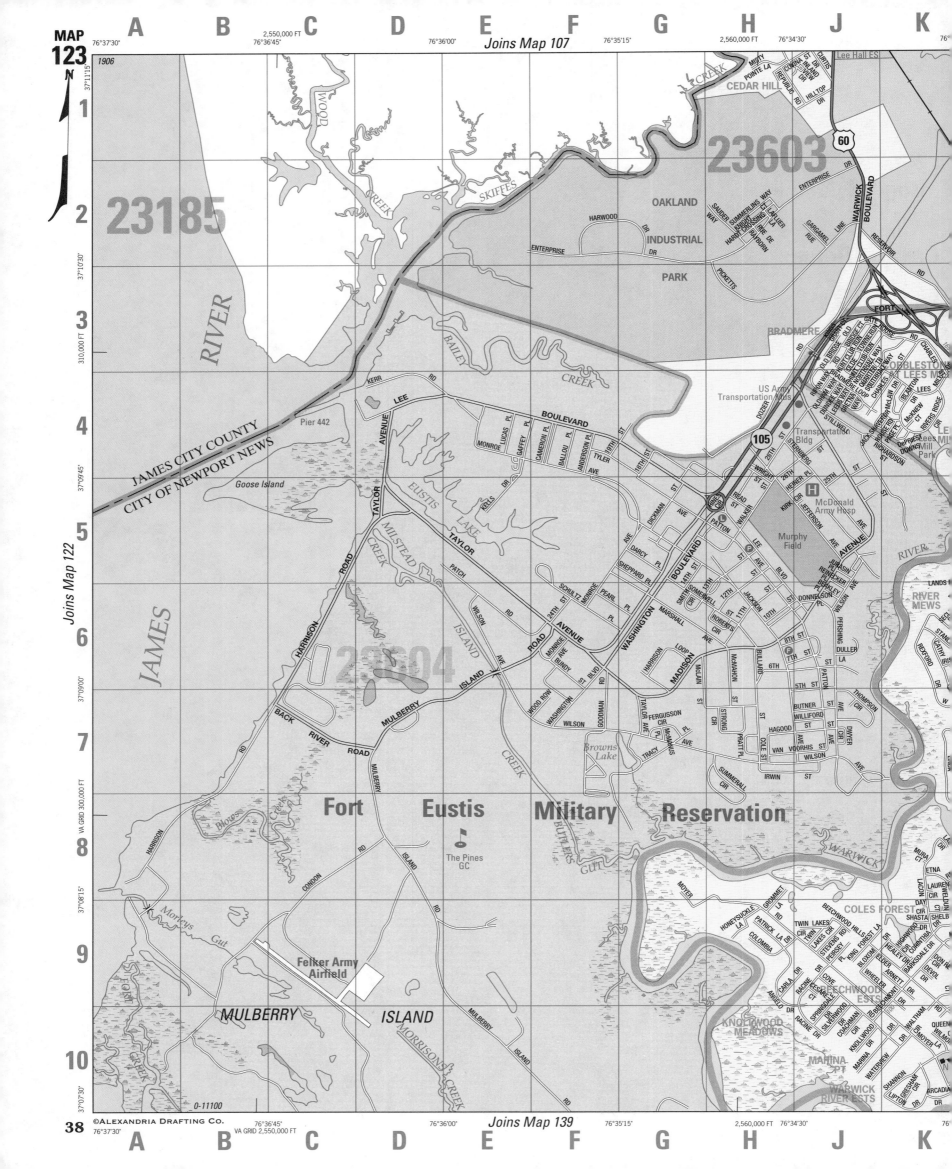

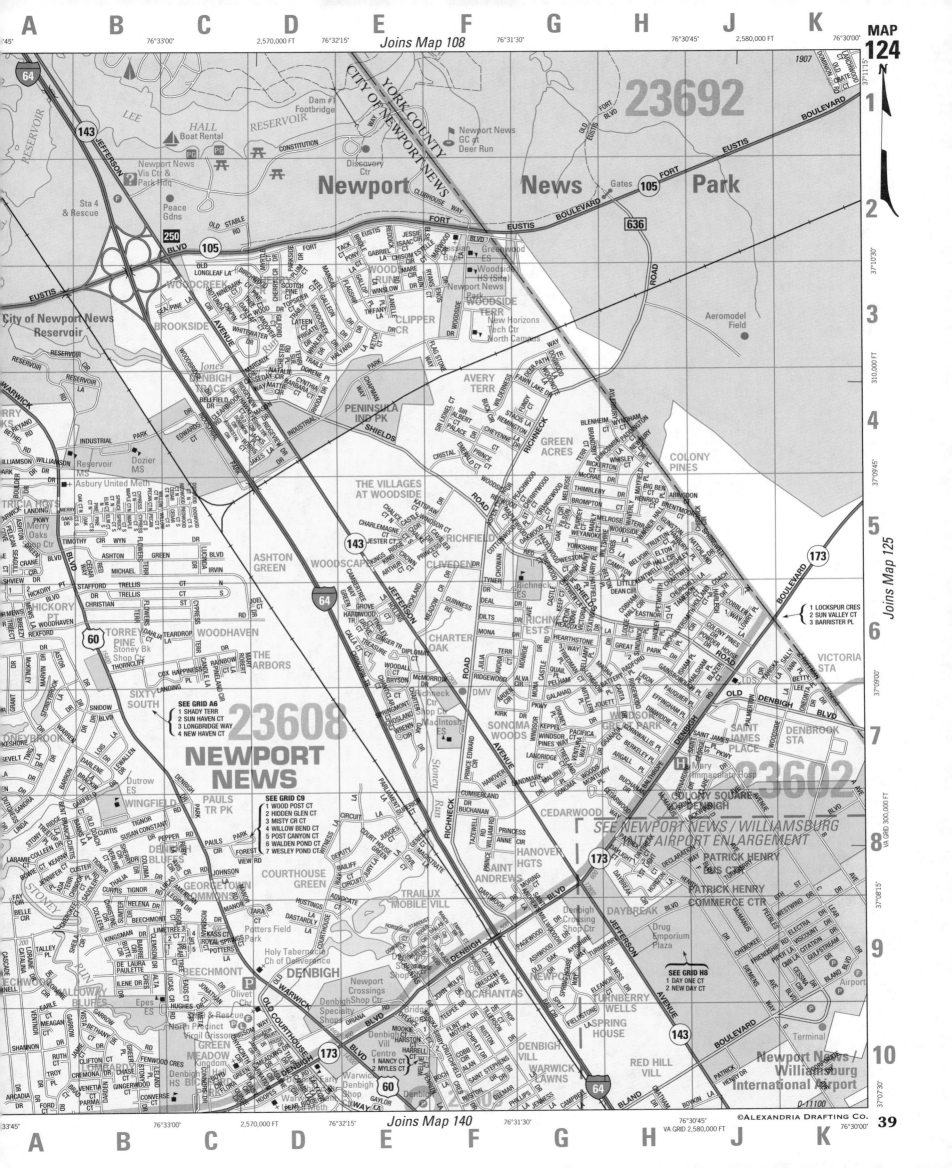

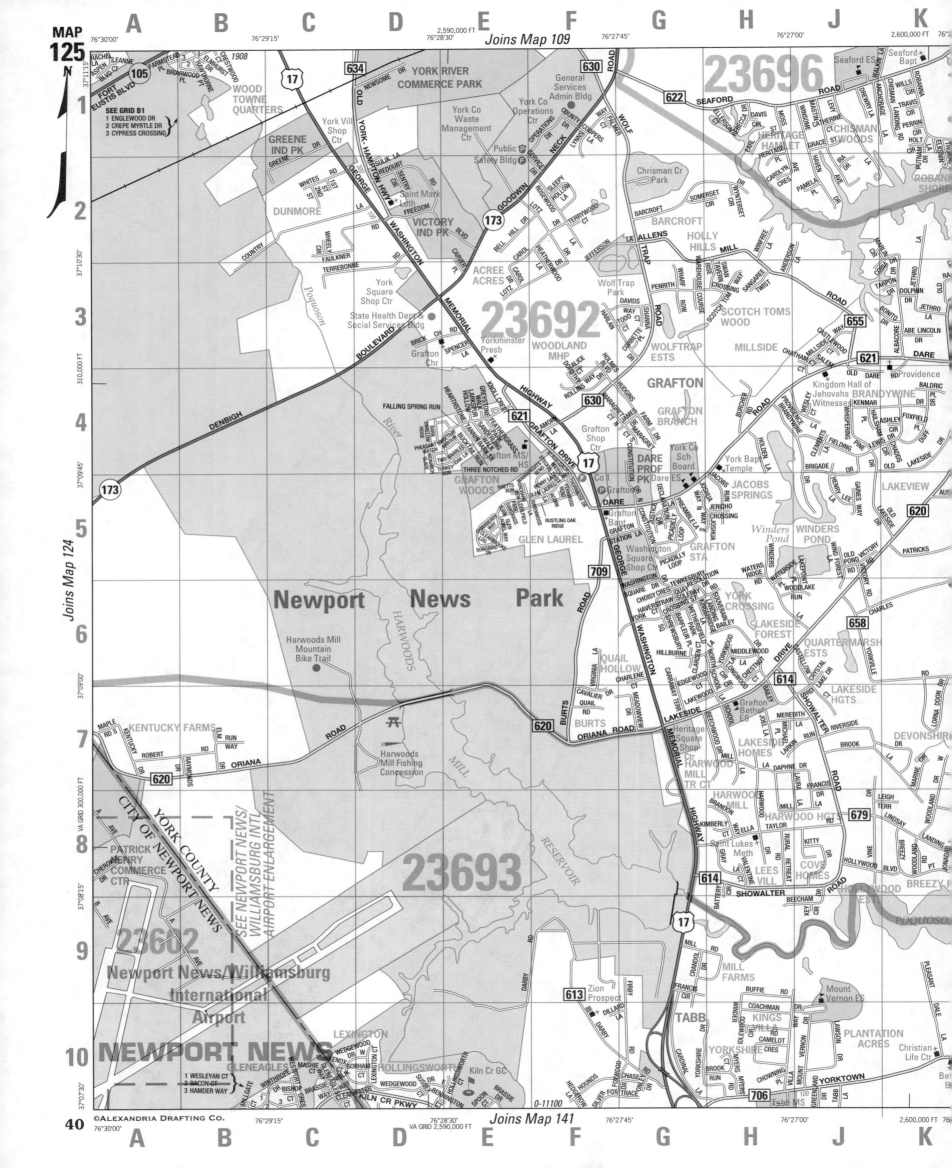

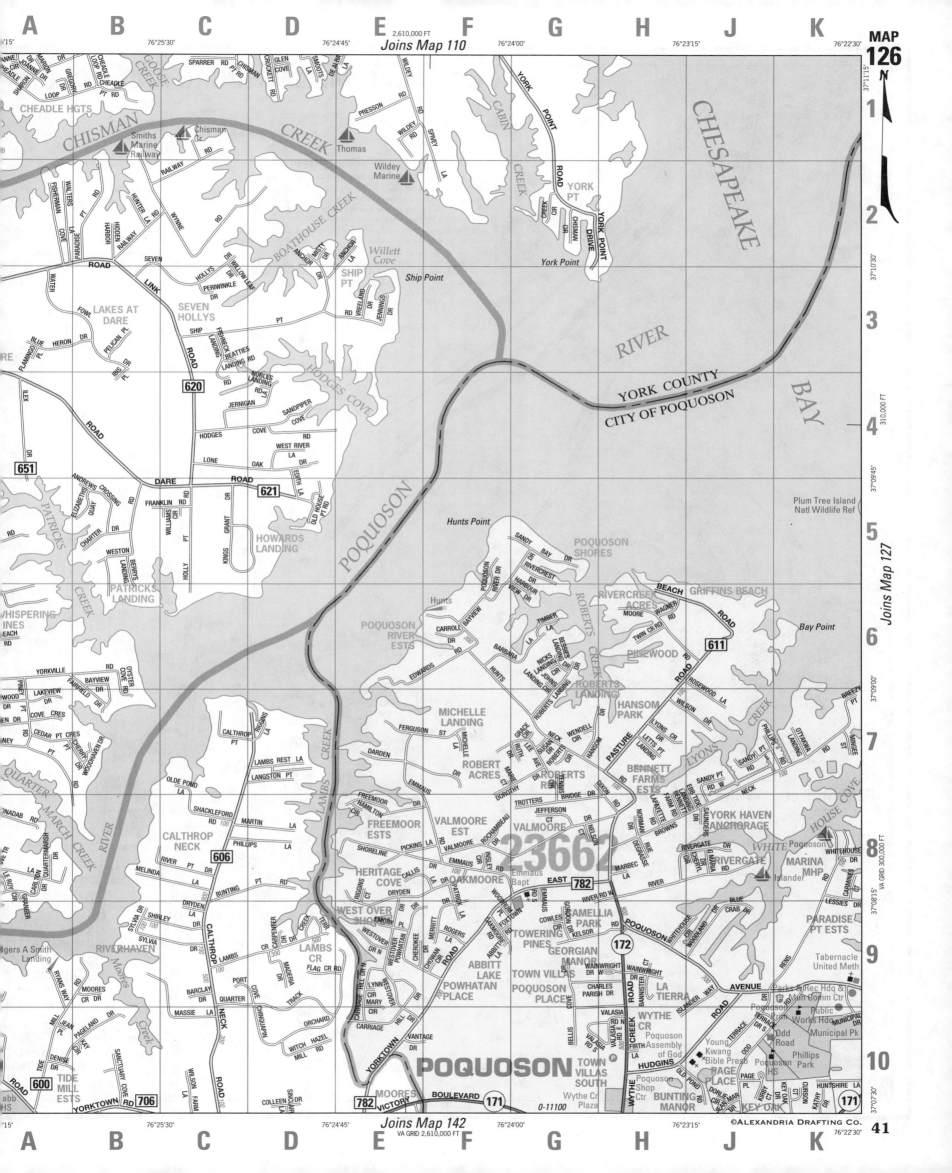

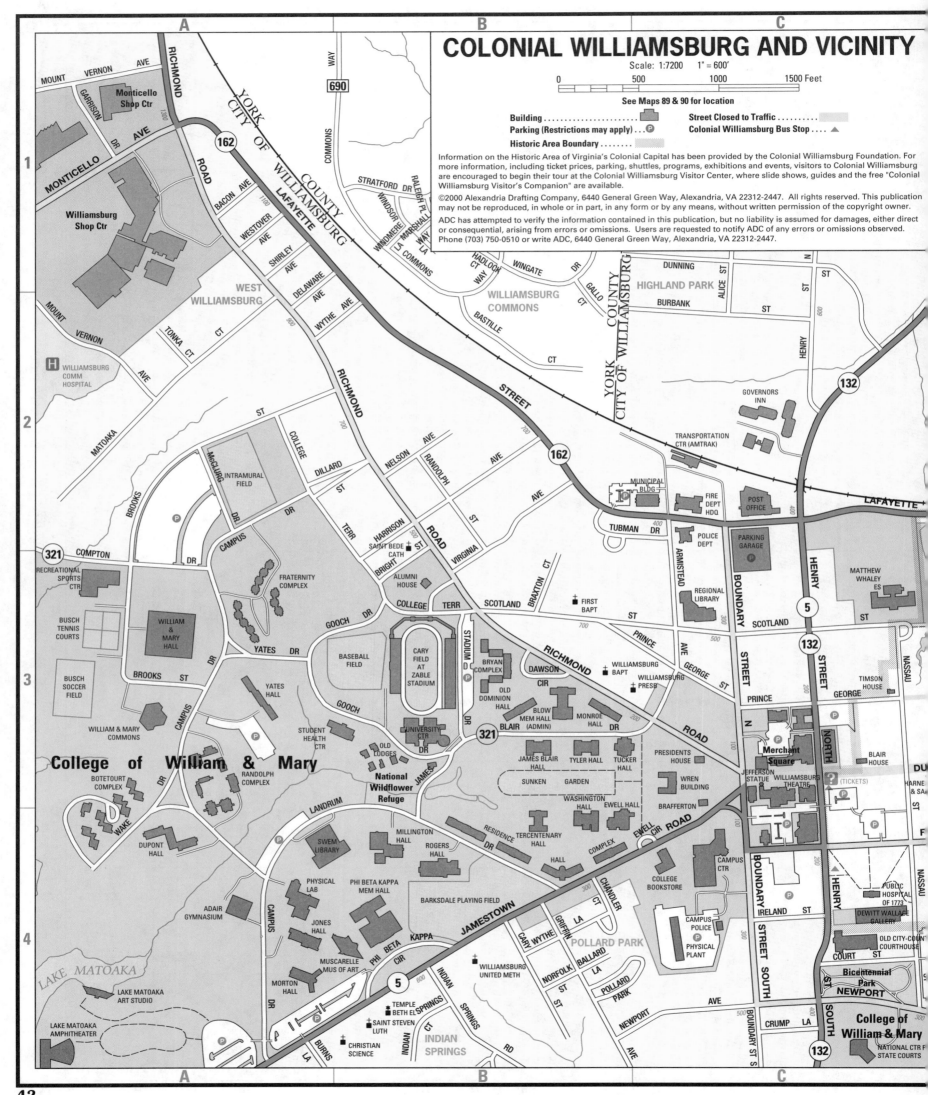

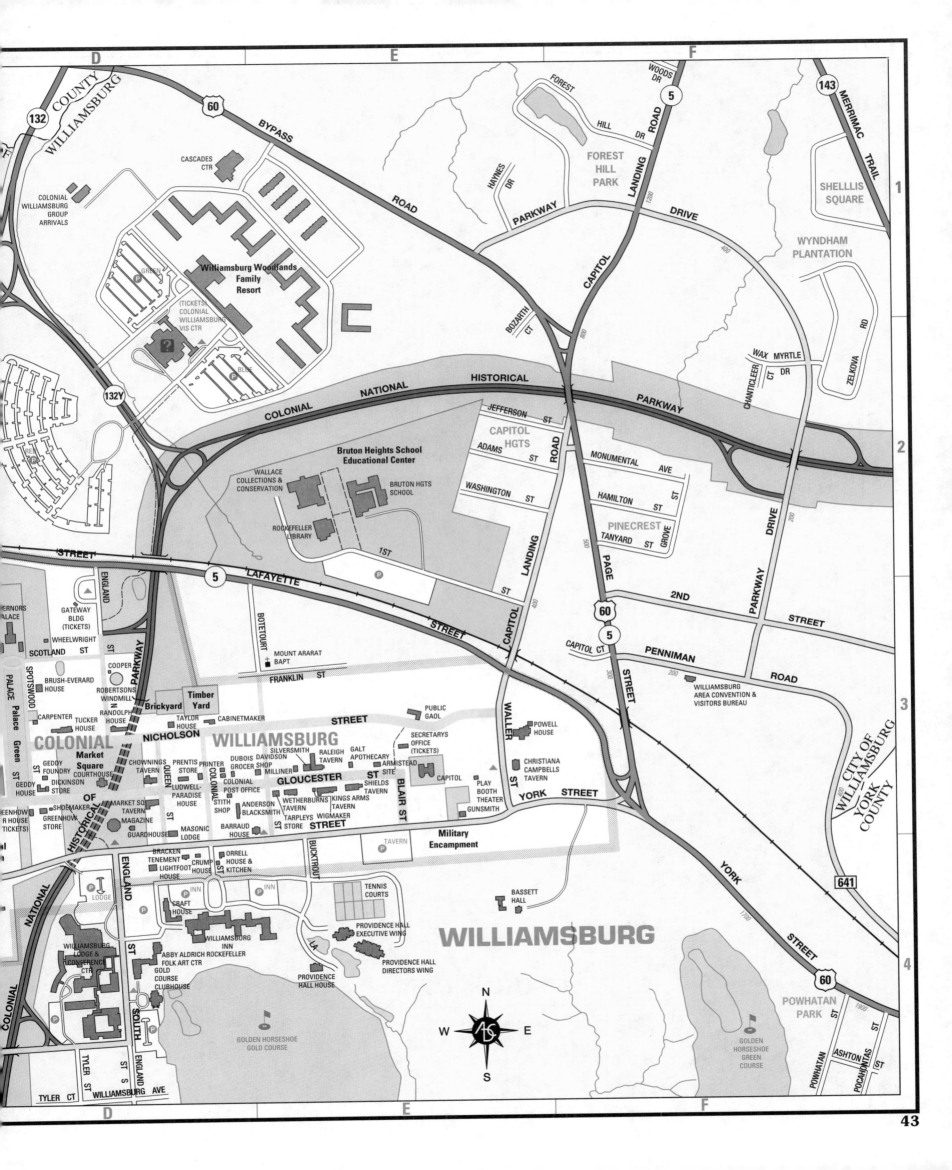

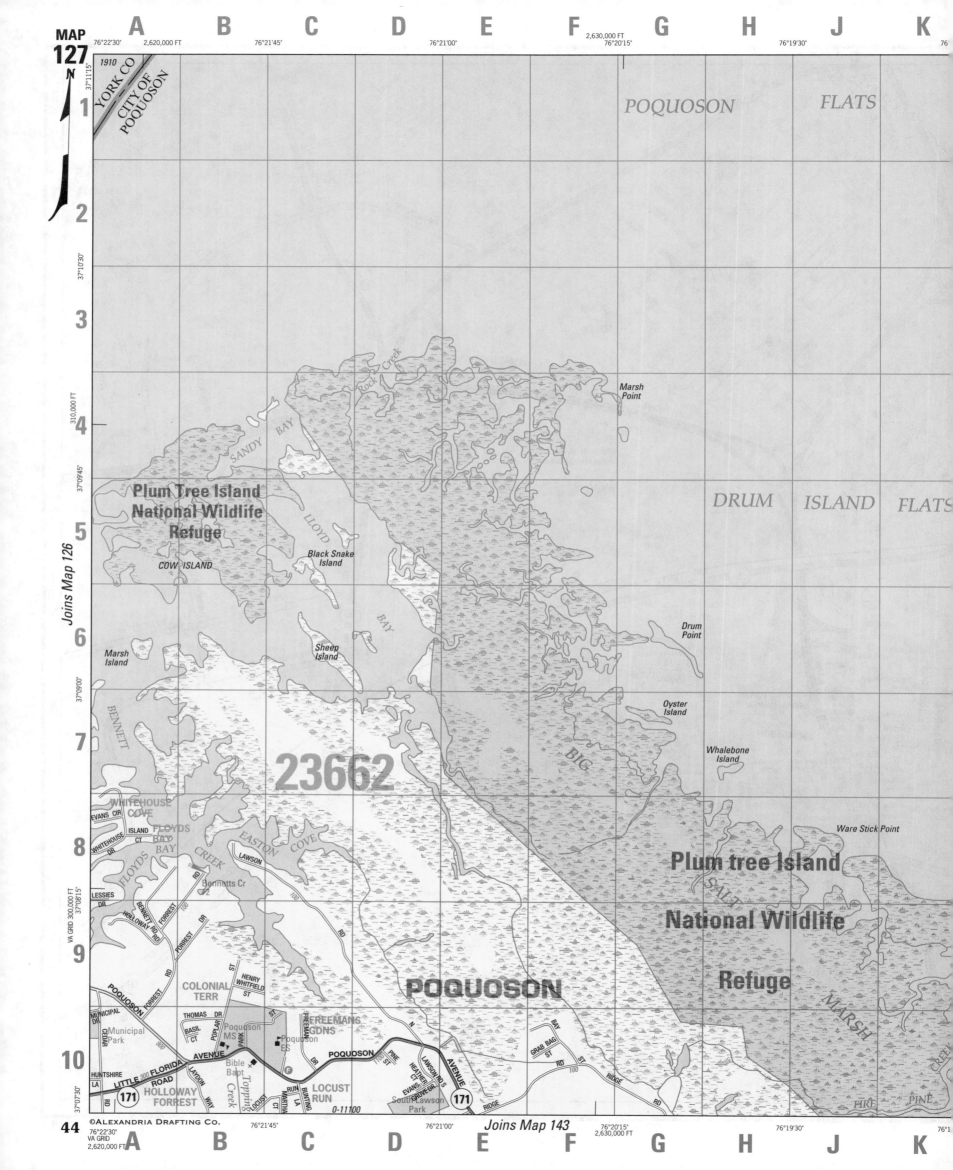

MAP
127

A B C D E F G H J K

76°22'30" 2,620,000 FT 76°21'45" 76°21'00" 2,630,000 FT 76°20'15" 76°19'30" 76

37°11'15"

1 1910 YORK CO CITY OF POQUOSON **POQUOSON** **FLATS**

2

37°10'30"

3

Rock Creek Marsh Point

37°10'00" 310,000 FT

4 SANDY BAY

Plum Tree Island LLOYD **DRUM** **ISLAND** **FLATS**

37°09'45"

National Wildlife

5 **Refuge** Black Snake Island

COW ISLAND BAY

Marsh Island Sheep Island Drum Point

6

37°09'00"

BENNETT Oyster Island

7 **23662** BIG Whalebone Island

WHITEHOUSE COVE

EVANS CIR Ware Stick Point

ISLAND FLOYDS

8 WHITEHOUSE DR BAY CREEK EASTON COVE **Plum tree Island**

37°08'15" VA GRID 300,000 FT FLOYDS BAY LAWSON SALT

LESSIES DR Bennetts Cr **National Wildlife**

BENNETT RD FORREST DR 100 RD 100

HOLLOWAY RD FORREST

9 ST **Refuge**

POQUOSON FORREST RD COLONIAL TERR HENRY HENRY WHITFIELD ST MARSH

MUNICIPAL DR THOMAS DR ST FREEMANS GDNS

CEDAR Municipal Park BASIL CT POPLAR Poquoson MS FREEMAN **POQUOSON** BAY GRAB BAG ST RD

10 AVENUE Poquoson ES **POQUOSON** PINE ST ST 100 RIDGE RD

37°07'30" HUNTSHIRE LA LITTLE 300 FLORIDA Bible Bapt DR AVENUE RD PINE

RD **171** HOLLOWAY FORREST LAYDON WAY Topping Creek LOCUST RUN LOCUST RUN MARTHA LA BUNTING South Lawson Park LAWSON RD S HEATHER CT EVANS GROVE DR **171** RIDGE FIRE PINE

0-11100

76°22'30" VA GRID 2,620,000 FT 76°21'45" 76°21'00" 76°20'15" 2,630,000 FT 76°19'30" 76°1

A B C D E F G H J K

Joins Map 126

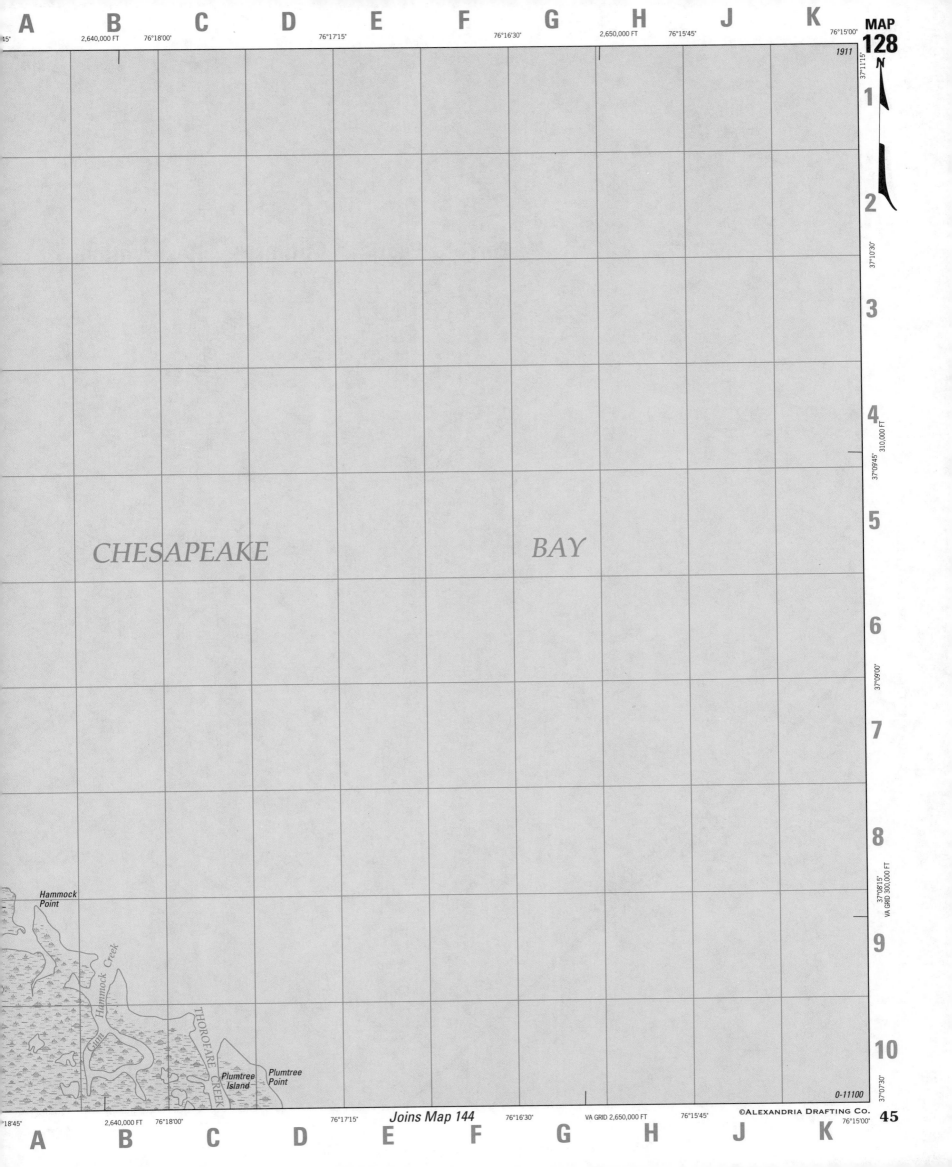

MAP
128

A B C D E F G H J K

1911

76°18'45"
2,640,000 FT
76°18'00"
76°17'15"
76°16'30"
2,650,000 FT
76°15'45"
76°15'00"

37°11'15"

1

2

37°10'30"

3

4

37°09'45"
310,000 FT

5

CHESAPEAKE   *BAY*

6

37°09'00"

7

8

37°08'15"
VA GRID 300,000 FT

Hammock Point

9

Hammock Creek

THOROFARE CREEK

10

Plumtree Island

Plumtree Point

0-11100

37°07'30"

76°18'45"
2,640,000 FT
76°18'00"
76°17'15"
Joins Map 144
76°16'30"
VA GRID 2,650,000 FT
76°15'45"
©Alexandria Drafting Co.
76°15'00"

45

A B C D E F G H J K

MAP
139

A B C D E F G H J K

76°37'30" 2,550,000 FT
76°36'45" 76°36'00" 76°35'15" 2,560,000 FT 76°34'30"

2018

N

37°07'30"

CREEK

MULBERRY

ISLAND

173

MULBERRY HILL

Denbigh Landing
Denbigh Park

HOOPE DANDIN

MORRISONS

Marshy Point

Fort   Eustis   Military   Reservation

ISLAND

Pauly Run

MULBERRY

RD

23604

Thorofare Point

Thorofare Island

THOROFARE

Swash Hole

SWASH

HOLE

ISLANDS

JAMES

37°06'45"
290,000 FT

37°06'00"

ISLAND

Crispy Point

VA GRID 280,000 FT
37°05'15"

RIVER

CITY OF NEWPORT NEWS
ISLE OF WIGHT COUNTY

37°04'30"

37°03'45"

0-11100

46

76°37'30" 76°36'45"
VA GRID 2,550,000 FT
76°36'00" 76°35'15" 2,560,000 FT 76°34'30"

A B C D E F G H J K

MAP
143
N

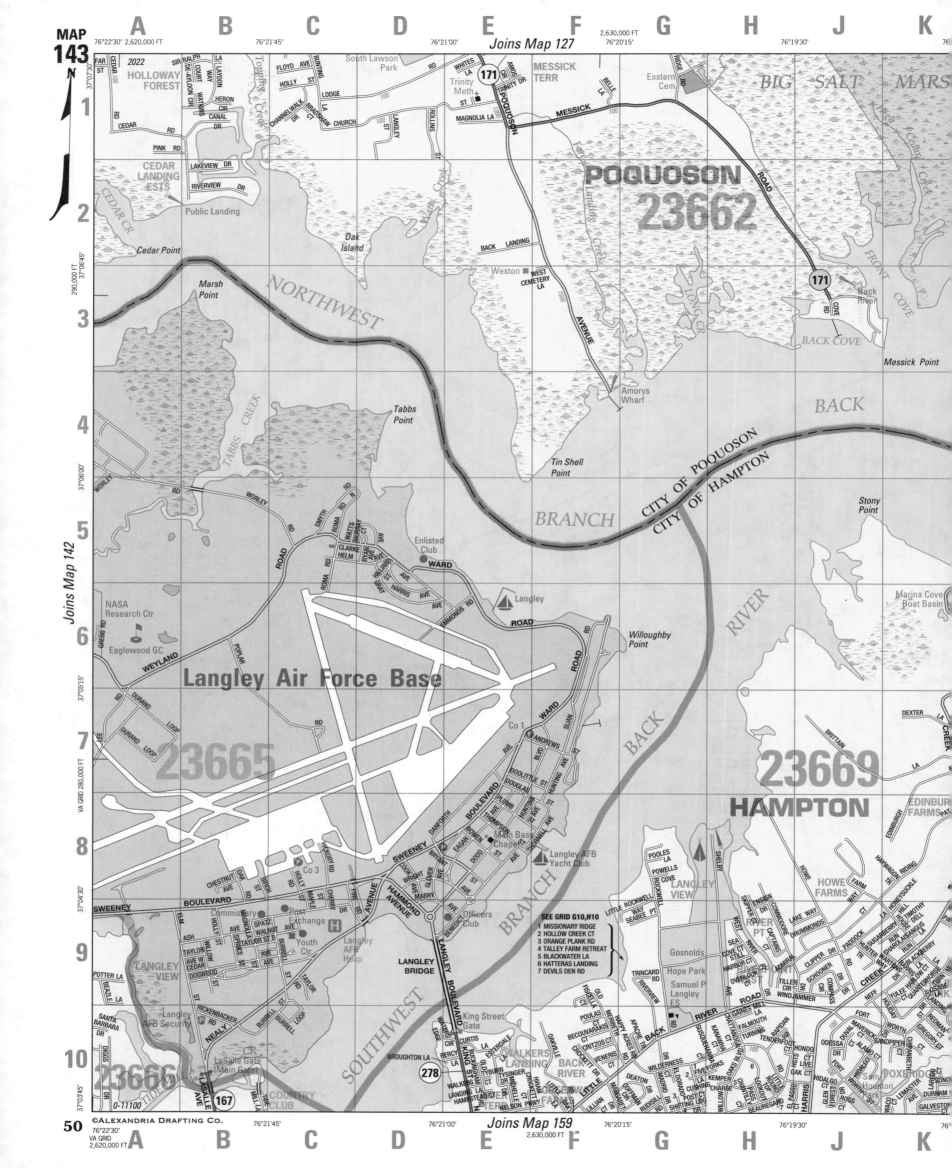

2022

HOLLOWAY FOREST

FAR CEDAR ST
SIR RALPH WAY
WATKINS
HERON CIR
LA LAYDON
COURT DR-AVLOON
CANAL DR
CEDAR RD
PINK RD
LAKEVIEW DR
RIVERVIEW DR

CEDAR LANDING ESTS

CEDAR CR

Public Landing

Cedar Point

Marsh Point

NORTHWEST

TABBS CREEK

Tabbs Point

Oak Island

Mill Creek

FLOYD AVE
BUNTING
HOLLY
LODGE
BRADSHAW
CT
HOLLY ST
CHURCH
LANGLEY ST
ROLLINS ST
CHANNELWALK DR

South Lawson Park

WHITES LA
AMOS CIR
TRINITY DR
Trinity Meth.
ST
MAGNOLIA LA

MESSICK TERR

171

POQUOSON

MESSICK

BELLE LA

Eastern Cem

RIDGE RD

Fords Landing Creek

POQUOSON
23662

BIG SALT MARS

Kings Cr

BACK LANDING

Weston
WEST CEMETERY LA

AVENUE

Amorys Wharf

LONG CR

FRONT COVE

171

COVE RD

Back River

Back COVE

Messick Point

BACK

BRANCH

CITY OF POQUOSON
CITY OF HAMPTON

Tin Shell Point

Stony Point

Willoughby Point

BACK

RIVER

Marina Cove Boat Basin

Joins Map 142
290,000 FT
37°06'45"
37°06'00"
37°05'15"
VA GRID 280,000 FT
37°04'30"
37°03'45"

WORLEY RD
WORLEY RD

SMYTH RD N
ROMA RD
WATTS AVE
MURRAY AVE
ROAD
CLARKE HELM
RYAN AVE
WILLARD ST
GRAY
HARRIS AVE
ROMA RD
S RD

ROAD

Enlisted Club
WARD

EMMONDS RD

Langley
ROAD

WARD RD

SLUNN ST

Co 1
ANDREWS

NASA Research Ctr

GREGG RD

Eaglewood GC

WEYLAND

POPLAR RD

Langley Air Force Base

23665

LEE RD
DURAND RD
DURAND LOOP

RD

23669

HAMPTON

DEXTER

BRITTAIN

CREEK LA

DOOLITTLE ST
DOUGLAS ST
PLUMB AVE
HUNTING AVE
THOMPSON
BOULEVARD
BUD
HUNTING AVE

EDINBURG FARMS

SWEENEY

DANFORTH

BRYANT
EAGAN AVE
DODD ST
THORNELL AVE
Main Base Chapel
BOWEN
WRIGHT

CHESTNUT AVE
OAK AVE
BIRCH ST
HICKORY RD
HOLLY ST
COOK RD

Co 3

Langley AFB Yacht Club

POOLES LA
POWELLS RD
ROCKWELL

Langley VIEW

HOWE FARMS

SHELBY

HARWAGON RIDING

HOWE FARM WAY

SWEENEY
BOULEVARD

ELM AVE
ASH
TAYLOR
WILLOW
AVE W CEDAR
DOGWOOD
SPRUCE ST
STAYLOR ST E
WALNUT AVE
SPATZ AVE
BURRELL

Commissary

Post Exchange

H
Youth Ctr

Langley AFB Hosp

THE AVENUE
MAPLE
CHERRY
HAMMOND

HAMMOND AVENUE

MABRY AVE

BENEDICT AVE

Officers Club

SEE GRID G10,H10
1 MISSIONARY RIDGE
2 HOLLOW CREEK CT
3 ORANGE PLANK RD
4 TALLEY FARM RETREAT
5 BLACKWATER LA
6 HATTERAS LANDING
7 DEVILS DEN RD

Little Rockwell WAY
SEABEE PT

Gosnolds
Hope Park

Samuel P Langley ES

TRINCARD RD

ENSIGN
COMMODORE
SEA COVE CT
HARBOR CT
TILLER CIR
WINDJAMMER

SCHOONER
COMPASS

RIVER PT

CAPTAINS

LAKE WAY

SKIPPER CT
ADMIRAL
OVERLOOK

SUGARBERRY
HUNTER WARTON RD
CHARLASANGUS
BLACKBERRY

POTTER LA
BEAZLE LA

LANGLEY VIEW

SANTA BARBARA DR

TAYLOR RD
BURRELL ST
BURRELL LOOP

RICKENBACKER
NEALY

Langley AFB Security

SOUTHWEST

LANGLEY BRIDGE

LANGLEY BOULEVARD

WATERS EDGE

King Street Gate

BRANCH

FISHELLA
OLD BUCKROE LA
POULAS
BECOUVARAKIS
CRITZOS
HAPPY ACRES
APACHE LA
OAKVILLE
RIVERVIEW

BACK

RIVER ROAD

GAINES
FALMOUTH
TURNING
RAPIDAN
TENDERFOOT
KANAM
SUSQUEHANNA

PADDOCK
YULE CT
CHAMBERTON
PINEWOOD
QUARTERHORSE
OSAGE
NEFF

HONEYSUCKLE
HILL
TIMOTHY
RUN

CREEK

FORT

23666

HIDEAWAY

DIGGS
LASALLE AVE
167
0-11100

LaSalle Gate (Main Gate)

CR
LITTLE JOHN

COUNTRY CLUB

BROUGHTON LA

278

KING ST
CURTIS
PERCY LA
LA

WALKERS LANDING
WALKER LANDING
HAMPSTEAD CT
NELSON
BUCKINGHAM CT
FALA

ESTERDALE
TYBURN
HIRAM

DEATON DR
RUDISILL
POSTING
SHIFTING LOG

EL DORADO
WILDERNESS
FIVE FORKS
CUSHING
KEMPER CT

MAVERICK
SANDPIPER
ODESSA
HONDO
EL PASO
RUTH

WORTH
GREGORY
ALAMO
LEE
CT
ROSS
GALVESTON

Sandy Houston Park

OXBRIDGE

HARRIS

HIDALGO CT
LOMASTER
LIVE OAK
CT

50
©ALEXANDRIA DRAFTING CO.
76°22'30"
VA GRID
2,620,000 FT
A   B   C   D   E   *Joins Map 159*   F   2,630,000 FT   G   H   J   K
76°21'45"   76°21'00"   76°20'15"   76°19'30"   76°
37°03'45"

MAP
155

N

A B C D E F G H J K

76°37'30"   2,550,000 FT   76°36'45"   76°36'00"   76°35'15"   2,560,000 FT   76°34'30"

37°03'45"
37°03'30"
37°03'00"
37°02'15"
260,000 FT
37°01'30"
37°00'45"
VA GRID 250,000 FT
37°00'00"

CITY OF NEWPORT NEWS
ISLE OF WIGHT COUNTY

JAMES

RIVER

2130

Fort Boykin
Hist Park

PILE

FORT BOYKIN TR

MORGARTS RD

ARTISTS WAY

BEARS CROSSING LA

BEACH

BARTONS LANDING

MORGARTS DR

MORGARTS BEACH LA

MORGARTS BEACH RD

FARM RD

RD

18000

18200

18400

MORGARTS BEACH

673

MORGARTS BEACH   17400

DAYS NECK RD

WENLEY CIR

DAYS PT

GURWEN

PROPER PL

DAYS PT LA

POINT

POINT LA

DAYS

LA

TORMENTORS LA

HAVLOW LA

TORMENTOR

LAKE

ROAD

18600

RD

WRONG RD

23430

WILLIAMS

Days
Point

673

673

HAVERTY LA

DAYS   17400

18000

POPLAR POINT LA

POINT

ROAD

18400

LA

PKWY

LENNY DR

NEST

EAGLE

MONETTE

MONETTE LA

TORMENTOR CREEK

CREEK

PAGAN   RIVER

PAGAN   RIVER

Goodwin
Point

JAMES RIVER HGTS

707

RAINBOW RD

BROWNS MARINA

BOLLING BLVD

0-14300

52

76°37'30"   76°36'45"   76°36'00"   76°35'15"   2,560,000 FT   76°34'30"

A B C D E F G H J K

23606

2131

BEVERLY
HILLS

FERGUSON
COVE
QUEENS
VALENTINE
CT
PAULA
MARIA DR
MEETING
RD
S JAMES
LANTERN
CIR
LANTERN
LANDING
PAGE
BRIDGE
RD
OAK RD
MUSHROOM
CT
INDIGO
DOWNING
PL

Lake
Queen
Anne

Indigo
Lake

DRAPER
LA

MERRY
VIEW

James River
CC

COUNTRY
CLUB
FARVIEW
RD
ROBERTS RD
RD
SCOTLAND

FAIRVIEW

Blunt
Point

JAMES
LANDING

RIVERSEDGE
RD
MADISON
CIR
LOOKOUT
CIR
MADISON
LA S
BLUFF RD
LITTLE
MADISON
MERRY
LA
MERRY PT
TERR
JACOBS ST
CHESTER
SPRING RD
SPOTTSWOOD
RD
LA
MALLICOTTE
LA
SMITH RD
CAPTAIN
JOHN

MERRY PT
ESTS

MCCROSS RD

RIVERSIDE
SHORE

SIR FRANCIS
WYATT RD
SELDEN
LA
BANDECOOT
CT

NEWPORT NEWS

RIVERSIDE
DR

RIVERSIDE
PARKWAY

JAMES

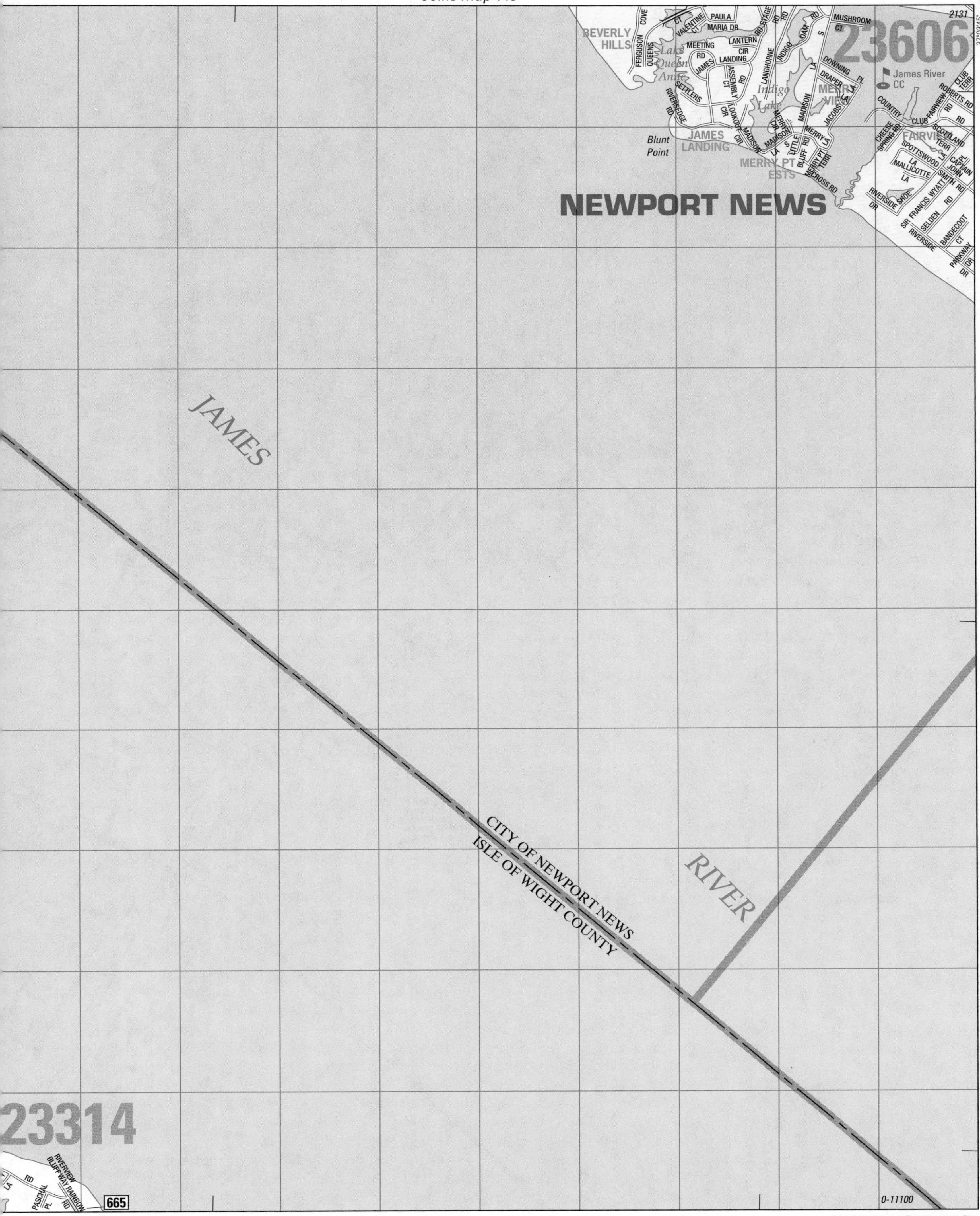

Joins Map 157

RIVER

CITY OF NEWPORT NEWS
ISLE OF WIGHT COUNTY

23314

RIVERVIEW
LA
PASCHAL
PL
RD
BLUFF WAY
RAMBLIN
RD

665

270,000 FT
37°03'45"
37°03'00"
37°02'30"
37°02'15"
260,000 FT
37°01'30"
37°00'45"
VA GRID 250,000 FT
37°00'00"
76°30'00"

1
2
3
4
5
6
7
8
9
10
53

0-11100

©ALEXANDRIA DRAFTING CO.

See ADC's "SUFFOLK & ISLE OF WIGHT COUNTY, VA STREET MAP" For Continuation

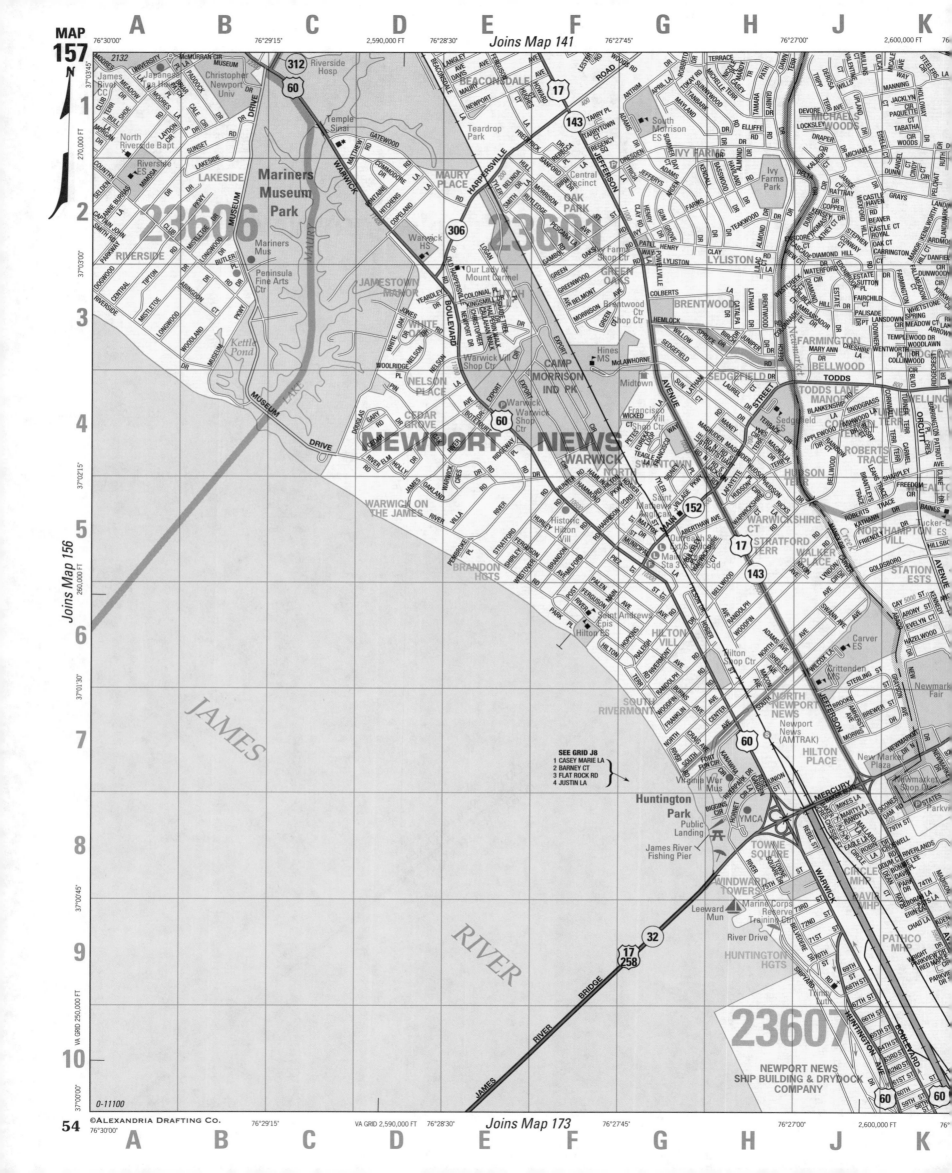

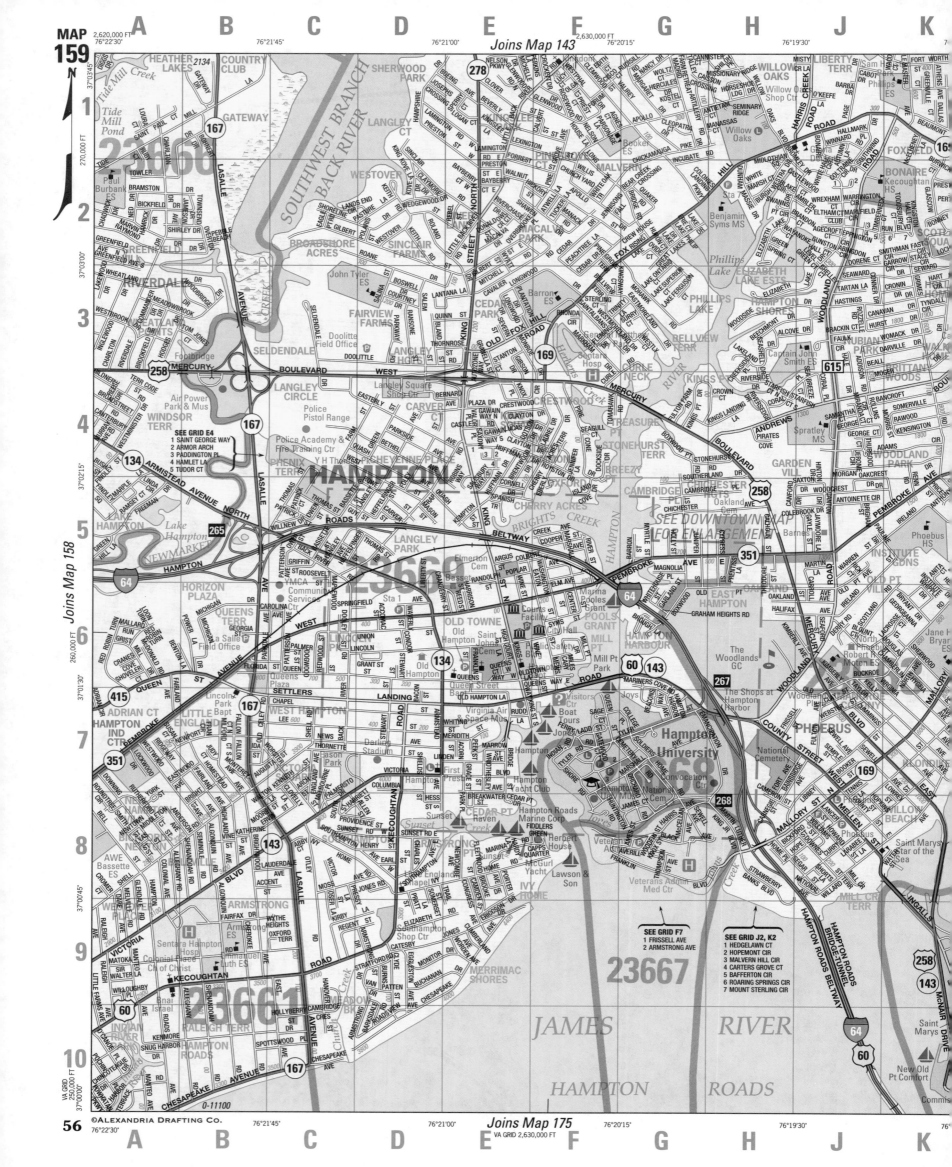

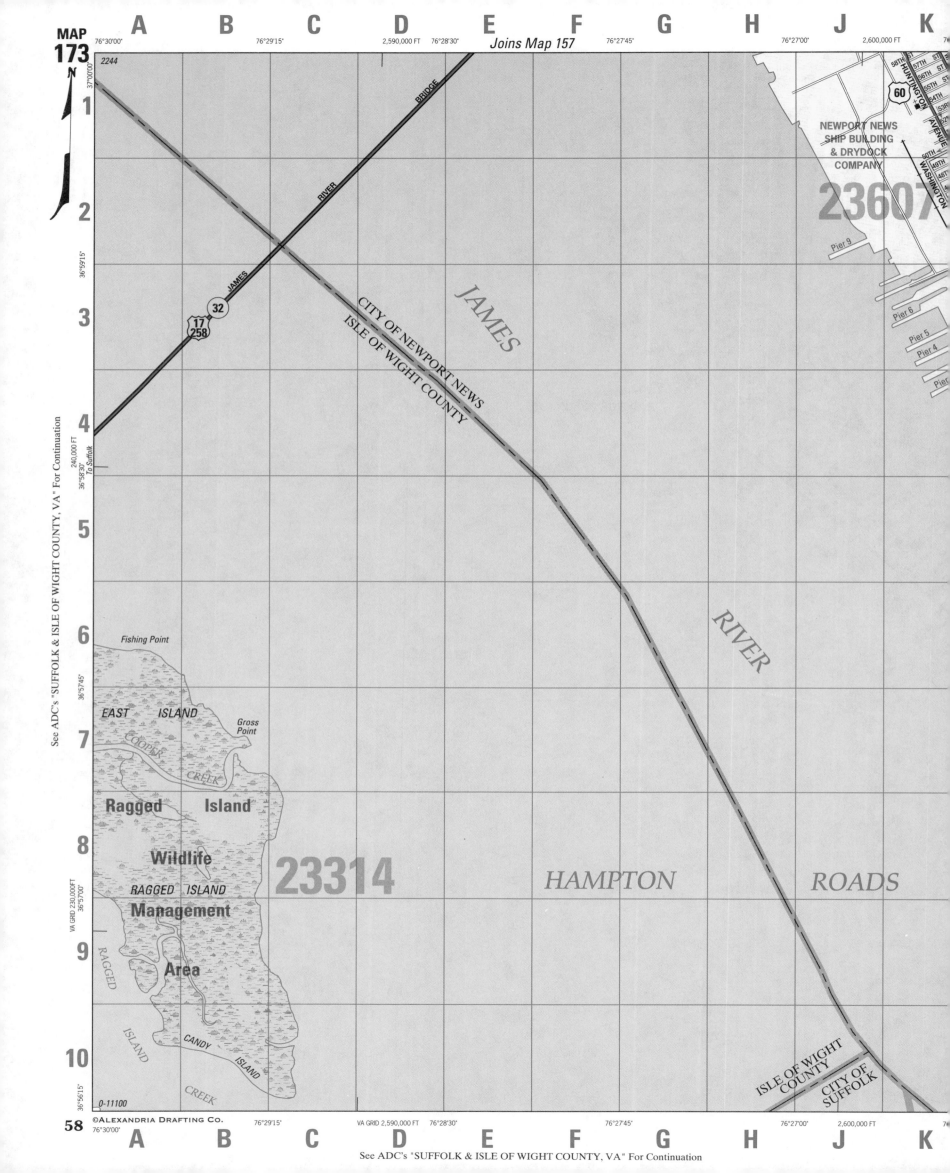

MAP
173

Joins Map 157

A **B** **C** **D** **E** **F** **G** **H** **J** **K**

76°30'00"   76°29'15"   2,590,000 FT   76°28'30"   76°27'45"   76°27'00"   2,600,000 FT

37°00'00"
2244

1

36°59'15"

2

36°58'30"

JAMES

RIVER

32

17
258

3

CITY OF NEWPORT NEWS

ISLE OF WIGHT COUNTY

JAMES

240,000 FT
To Suffolk

4

NEWPORT NEWS
SHIP BUILDING
& DRYDOCK
COMPANY

23607

Pier 9

Pier 6

Pier 5

Pier 4

Pier

60

58TH ST
57TH ST
56TH ST
55TH ST
53RD
HUNTINGTON
AVENUE
60TH
49TH
48TH
WASHINGTON
52

5

RIVER

6

Fishing Point

36°57'45"

EAST   ISLAND

7

COOPER

Gross
Point

CREEK

Ragged   Island

36°57'00"

8

Wildlife

RAGGED   ISLAND

VA GRID 230,000FT

Management

23314

HAMPTON   ROADS

9

RAGGED

Area

ISLAND

36°56'15"

CANDY   ISLAND

10

ISLE OF WIGHT
COUNTY

CITY OF
SUFFOLK

CREEK

0-11100

©ALEXANDRIA DRAFTING CO.

76°30'00"   76°29'15"   VA GRID 2,590,000 FT   76°28'30"   76°27'45"   76°27'00"   2,600,000 FT

A **B** **C** **D** **E** **F** **G** **H** **J** **K**

See ADC's "SUFFOLK & ISLE OF WIGHT COUNTY, VA" For Continuation

See ADC's "SUFFOLK & ISLE OF WIGHT COUNTY, VA" For Continuation

HAMPTON

NEWPORT
NEWS
23607

23605

23661

NEWSOME
PARK

CITY OF HAMPTON
CITY OF NEWPORT NEWS

EAST END

STUART GDNS

CHRISTOPHERS
SHORES

Anderson
Park

Monitor-Merrimac
Overlook Park

NEWPORT NEWS
MARINE
TERMINAL

Newport News
Cruise
Terminal

LASSITER
CT

Newport News
Tour Boat

King Lincoln
Park

SEAFOOD
IND PK

Virginia Marine
Resource
Commission

Newport
News Point

ROADS

HAMPTON

MONITOR-MERRIMAC
HAMPTON
ROADS
MEMORIAL
BRIDGE-TUNNEL
BELTWAY

CITY OF NEWPORT NEWS
CITY OF NORFOLK

CITY OF PORTSMOUTH
CITY OF NORFOLK

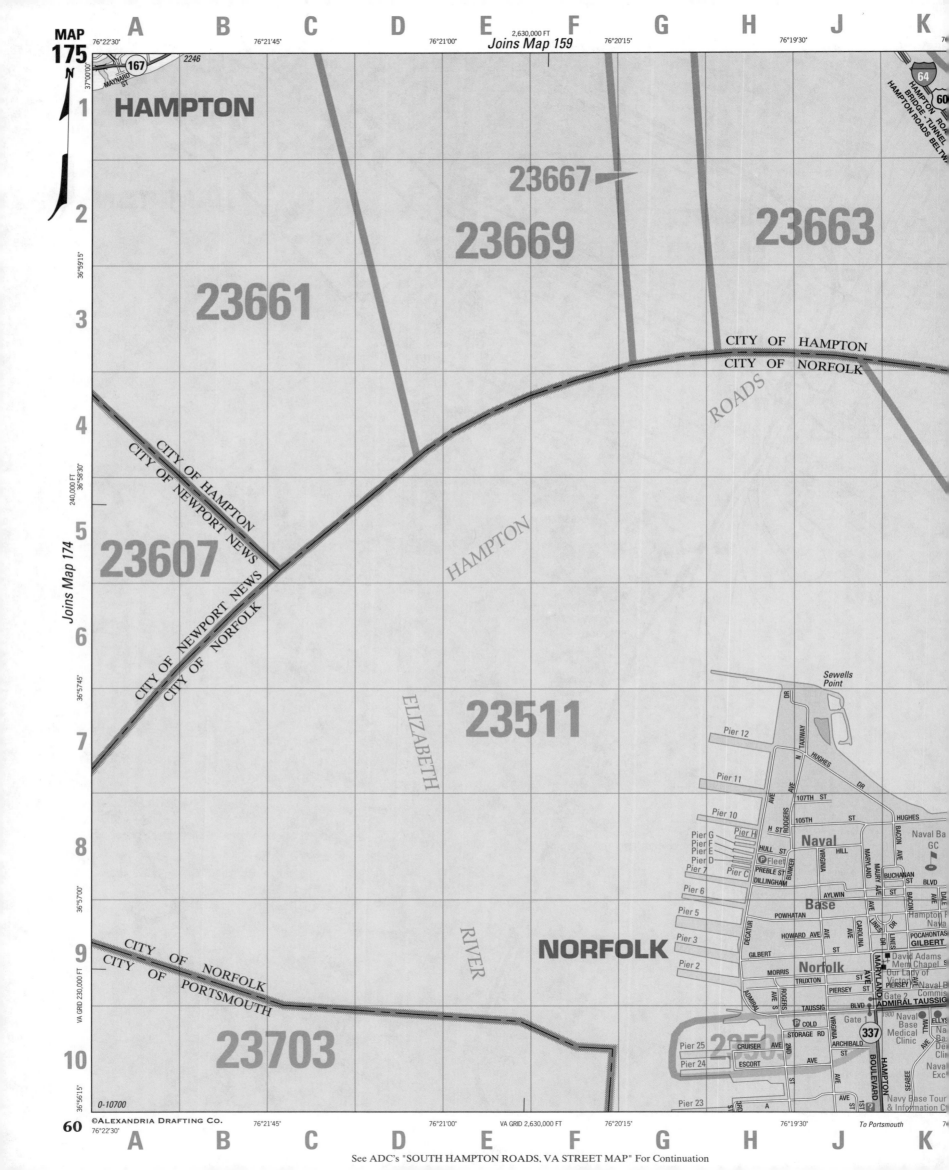

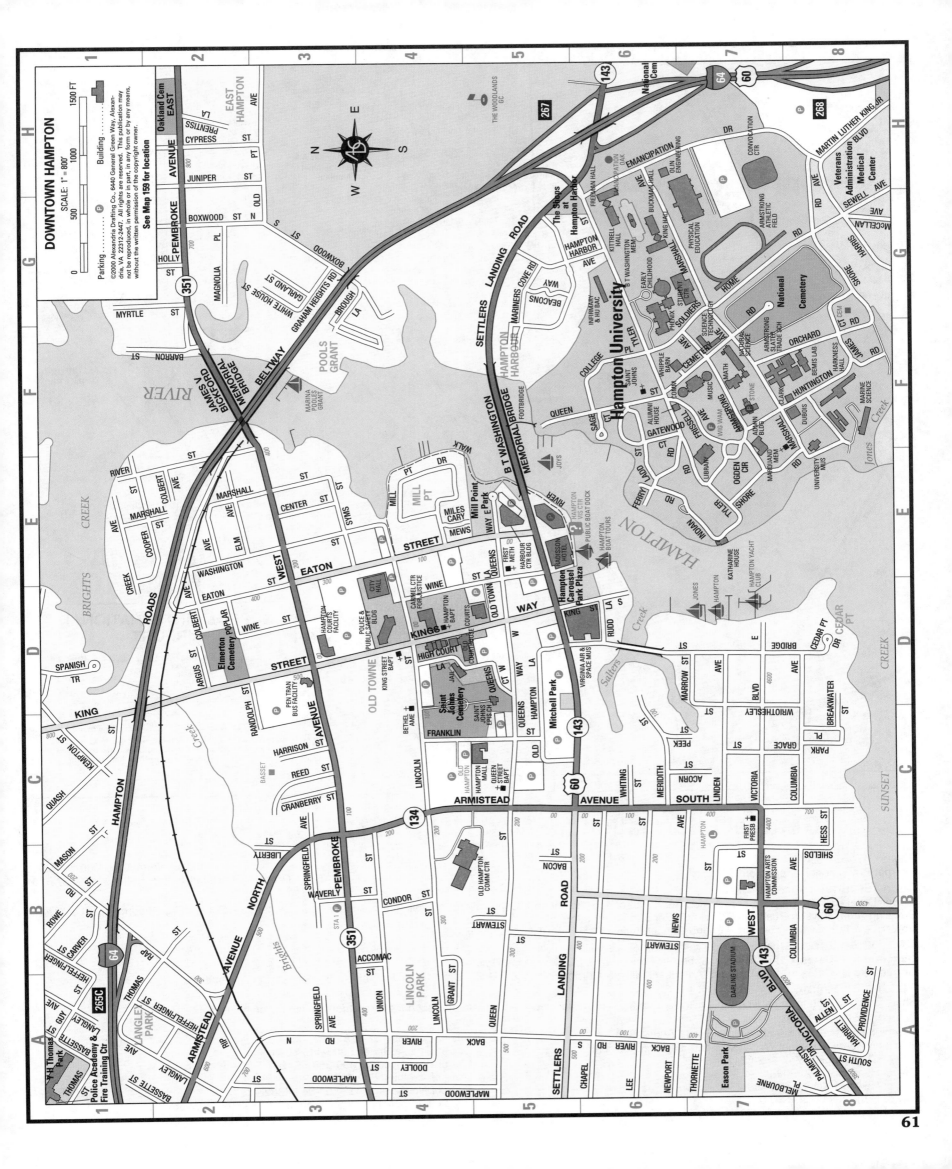

61

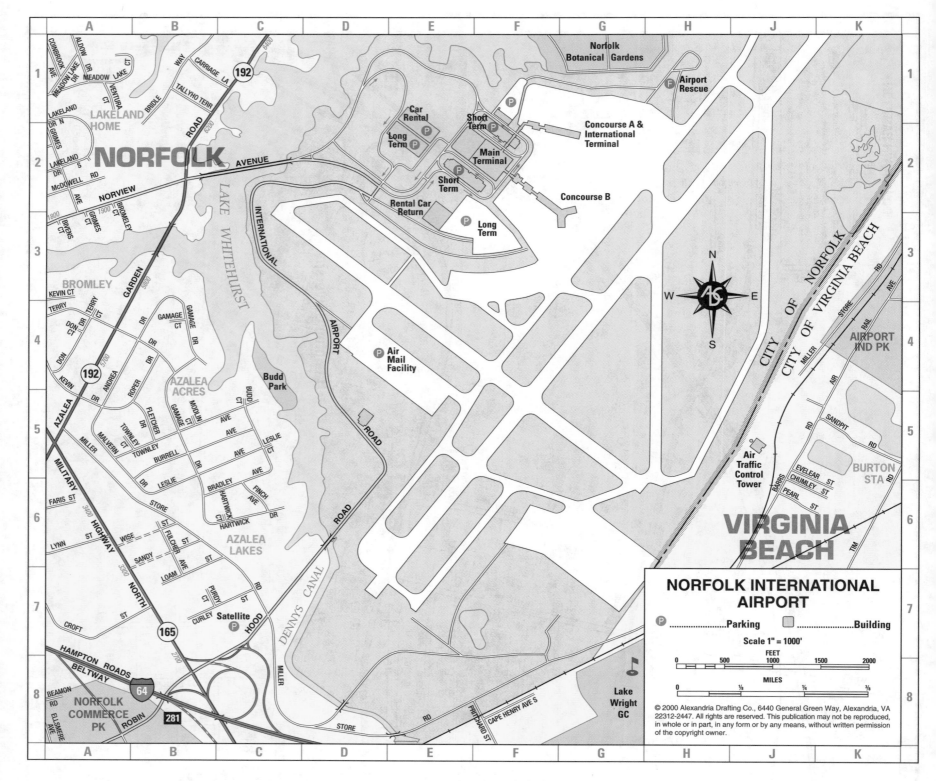

General Information for Norfolk International Airport (ORF)

Phone: 757-857-3200
Internet: www.norfolkairport.com

Traffic Information: 800-792-2800
Groome Chartered Transportation: 757-857-1231

Main Terminal Information:

Upper Level - ATM's, Barber Shop, Armed Services Lounge, Restaurant, Concessions, Norfolk International Airport Conference Center, Flight Information, Lockers, Smoking Area, Travel & Business Service Center and Visitor Information.

Lower Level - Passenger pick-up, Ground Transportation, Baggage Claim and Rental Cars.

Concourse Information:

Concourse A (Gates 1-15) Airlines: Cape Air, Trans States, USAirways and TWA.

Concourse B (Gates 16-30) Airlines: American, Midway Corporate Airlines, Delta, United, Continental, Northwest and Shuttle America.

Parking Information:

Short term and long term parking are available for a fee. Follow detailed signs as you approach the airport for parking lot locations.

Tourist Information:

Hampton Roads Chamber of Commerce: 757-622-2312
Hampton Roads Regional Website: www.hamptonroads.com
City of Chesapeake: 757-382-6241
City of Hampton: 800-800-2202
City of Newport News: 888-493-7386
City of Norfolk: 800-368-3097
City of Portsmouth: 800-767-8782

City of Suffolk: 757-925-6339
City of Virginia Beach: 800-446-8038
City of Williamsburg: 800-368-6511
State of Virginia: 800-847-4882
State of North Carolina: 800-847-4862
Norfolk International Airport Conference Center: 757-857-3381

INDEX

Pages 2, 3 and 4 provide you a complete Step-by-Step Guide to using your Street Map Book, Map Legend, Table of Contents and Key to Abbreviations. Take a few minutes to familiarize yourself with this time saving information.

© Alexandria Drafting Co.

ZIP CODE AREAS SHOWN IN THIS PUBLICATION

Alphabetical Listing by Post Office

POST OFFICE	ZIP CODE	MAP
Barhamsville, New Kent Co., VA	23011	39,40,56
Carrollton, Isle of Wight Co., VA	23314	155,156,173
Charles City, Charles City Co., VA	23030	54,70,71,86,87
Denbigh, City of Newport News, VA	23609	124
Gloucester, Gloucester Co., VA	23061	41,57,58,74,91
Gloucester Point, Gloucester Co., VA	23062	92,108,109
Hampton, City of Hampton, VA	23670	158
Hampton (Buckroe Beach), City of Hampton, VA	23664	144,160
Hampton (Fort Monroe), City of Hampton, VA	23651	159,160,175
Hampton (Hampton University), City of Hampton, VA	23668	143
Hampton (Langley AFB), City of Hampton, VA	23665	142,143,159
Hampton (Old Hampton), City of Hampton, VA	23669	143,144,158-160,175
Hampton (Phoebus), City of Hampton, VA	23663	159,160,175
Hampton (Riverdale), City of Hampton, VA	23666	141-143,157-159
Hampton (Veterans Admin), City of Hampton, VA	23667	143
Hampton (Veterans Administration), City of Hampton, VA	23667	175
Hampton (Wythe), City of Hampton, VA	23661	158,159,174,175
Hayes, Gloucester Co., VA	23072	91,92,109,110
Lackey, York Co., VA	23694	108
Lanexa, New Kent Co., VA	23089	38,39,54,55,70,71
Lightfoot, York Co., VA	23090	72
Newport News, City of Newport News, VA	23607	157,158,173-175
Newport News (Denbigh), City of Newport News, VA	23602	124,125,139-142
Newport News (Denbigh), City of Newport News, VA	23608	123,124,139,140
Newport News (Fort Eustis), City of Newport News, VA	23604	122,123,139,140
Newport News (Hidenwood), City of Newport News, VA	23606	140,141,155-157
Newport News (Lee Hall), City of Newport News, VA	23603	107,108,123,124
Newport News (Oyster Point), City of Newport News, VA	23612	141
Newport News (Parkview), City of Newport News, VA	23605	157,158,174
Newport News (Warwick), City of Newport News, VA	23601	141,157
Norfolk (Navy Base/Fleet), City of Norfolk, VA	23511	174,175
Norfolk (Ocean View), City of Norfolk, VA	23503	175
Norfolk (Wright), City of Norfolk, VA	23505	175
Norge, James City Co., VA	23127	72
Poquoson, City of Poquoson, VA	23662	126-128,142-144
Portsmouth (Churchland), City of Portsmouth, VA	23703	174,175
Seaford, York Co., VA	23696	109,110,125,126
Shacklefords, King and Queen Co., VA	23156	40,41,57
Smithfield, Isle of Wight Co., VA	23430	122,139,155
Spring Grove, Surry Co., VA	23881	103,104
Suffolk (Crittenden), City of Suffolk, VA	23433	173
Suffolk (Driver), City of Suffolk, VA	23435	173
Surry, Surry Co., VA	23883	104-106,122
Toano, James City Co., VA	23168	39,40,55,56,70-72
Wicomico, Gloucester Co., VA	23184	92
Williamsburg, City of Williamsburg, VA	23187	89
Williamsburg, York Co., VA	23185	86-91,103-107,122,123
Williamsburg, York Co., VA	23188	40,41,56-58,70-74,86-91,104
Yorktown, York Co., VA	23690	107-109
Yorktown, York Co., VA	23693	124-126,141,142
Yorktown (Grafton), York Co., VA	23692	108-110,124-126
Yorktown (US Naval Weapons Station), York Co., VA	23691	90-92,106-108

Numerical Listing by ZIP Code

ZIP CODE	POST OFFICE	MAP
23011	Barhamsville, New Kent Co., VA	39,40,56
23030	Charles City, Charles City Co., VA	54,70,71,86,87
23061	Gloucester, Gloucester Co., VA	41,57,58,74,91
23062	Gloucester Point, Gloucester Co., VA	92,108,109
23072	Hayes, Gloucester Co., VA	91,92,109,110
23089	Lanexa, New Kent Co., VA	38,39,54,55,70,71
23090	Lightfoot, York Co., VA	72
23127	Norge, James City Co., VA	72
23156	Shacklefords, King and Queen Co., VA	40,41,57
23168	Toano, James City Co., VA	39,40,55,56,70-72
23184	Wicomico, Gloucester Co., VA	92
23185	Williamsburg, York Co., VA	86-91,103-107,122,123
23187	Williamsburg, City of Williamsburg, VA	89
23188	Williamsburg, York Co., VA	40,41,56-58,70-74,86-91,104
23314	Carrollton, Isle of Wight Co., VA	155,156,173
23430	Smithfield, Isle of Wight Co., VA	122,139,155
23433	Suffolk (Crittenden), City of Suffolk, VA	173
23435	Suffolk (Driver), City of Suffolk, VA	173
23503	Norfolk (Ocean View), City of Norfolk, VA	175
23505	Norfolk (Wright), City of Norfolk, VA	175
23511	Norfolk (Navy Base/Fleet), City of Norfolk, VA	174,175
23601	Newport News (Warwick), City of Newport News, VA	141,157
23602	Newport News (Denbigh), City of Newport News, VA	124,125,139-142
23603	Newport News (Lee Hall), City of Newport News, VA	107,108,123,124
23604	Newport News (Fort Eustis), City of Newport News, VA	122,123,139,140
23605	Newport News (Parkview), City of Newport News, VA	157,158,174
23606	Newport News (Hidenwood), City of Newport News, VA	140,141,155-157
23607	Newport News, City of Newport News, VA	157,158,173-175
23608	Newport News (Denbigh), City of Newport News, VA	123,124,139,140
23609	Denbigh, City of Newport News, VA	124
23612	Newport News (Oyster Point), City of Newport News, VA	141
23651	Hampton (Fort Monroe), City of Hampton, VA	159,160,175
23661	Hampton (Wythe), City of Hampton, VA	158,159,174,175
23662	Poquoson, City of Poquoson, VA	126-128,142-144
23663	Hampton (Phoebus), City of Hampton, VA	159,160,175
23664	Hampton (Buckroe Beach), City of Hampton, VA	144,160
23665	Hampton (Langley AFB), City of Hampton, VA	142,143,159
23666	Hampton (Riverdale), City of Hampton, VA	141-143,157-159
23667	Hampton (Veterans Admin), City of Hampton, VA	143
23667	Hampton (Veterans Administration), City of Hampton, VA	175
23668	Hampton (Hampton University), City of Hampton, VA	143
23669	Hampton (Old Hampton), City of Hampton, VA	143,144,158-160,175
23670	Hampton, City of Hampton, VA	158
23690	Yorktown, York Co., VA	107-109
23691	Yorktown (US Naval Weapons Station), York Co., VA	90-92,106-108
23692	Yorktown (Grafton), York Co., VA	108-110,124-126
23693	Yorktown, York Co., VA	124-126,141,142
23694	Lackey, York Co., VA	108
23696	Seaford, York Co., VA	109,110,125,126
23703	Portsmouth (Churchland), City of Portsmouth, VA	174,175
23881	Spring Grove, Surry Co., VA	103,104
23883	Surry, Surry Co., VA	104-106,122

EXPLANATION OF THE ABOVE CHART:

Example: Yorktown (Grafton) Post Office is physically located in York County and serves ZIP Code Area 23692. ZIP Code Area 23692 is shown on maps 108-110, 124-126 of this publication.

TERRITORY CODES

Use the code letters behind each name and location to find the vicinity in which the street is located. The codes are as follows:

CC .... Charles City County, VA
F .... Fort Eustis, VA
FM .... Fort Monroe, VA
G .... Gloucester County, VA
H .... Hampton, VA
I .... Isle of Wight County, VA
J .... James City County, VA
KQ .... King and Queen County, VA
L .... Langley AFB, VA
N .... Norfolk, VA
NBN .... Naval Base Norfolk, VA
NK .... New Kent County, VA
NN .... Newport News, VA
P .... Portsmouth, VA
Pq .... Poquoson, VA
SC .... Surry County, VA
W .... Williamsburg, VA
Y .... York County, VA

Example: Abby Ct 159-C8 H would be on map 159, grid square C8 in Hampton.

Example: Bastille Ct WE-B2 Y would be on Colonial Williamsburg Enlargement map, grid square B2 in York County.

Example: Cemetery Rd HE-F7 H would be on Hampton Enlargement map, grid square H7 in Hampton.

STREETS

4-H Club Rd 104-D4, F3 J

A

A Ave 175-H10 N; 175-H10 NBN; 124-K7; 125-A8, A9 NN; 109-J6 Y
A St 91-B5 Y
Abaco Dr 157-K6 H
Abbey Ct 140-E4 NN
Abbitt La 104-J5 J; 140-J5 NN; 126-F9 Pq
Abbotsford Mews 89-B3 J
Abby Ct 159-C8 H
Abe Lincoln Dr 125-K3 Y
Aberdeen 88-K3 J
Aberdeen Ave 174-E1 NN
Aberdeen Rd 158-E4, F9, F10; 174-F1 H
Aberfeldy Way 141-F1 Y
Aberthaw Ave 157-G5 NN
Abigale La 106-B3 J
Abingdon Cir 159-J2 H
Abingdon Ct 124-H5; 157-B3 NN
Abingdon Hgts Dr 92-J4 G
Abraham Ct 158-C7 NN
Acacia Ct 105-G2 J
Academy La 141-C4 NN
Accent St 159-B8 H
Accomac St 159-D6; HE-B3 H
Accomac Turning 142-D1 Y
Accorn Ct 72-F1 J
Achievement Way 141-D5 NN
Acorn Ave 174-E1 NN
Acorn La 125-F5 Y
Acorn St 159-E7; HE-C7 H
Acton Dr 140-D4 NN
Ada Terr 124-A8 NN
Adams Ave 157-H6 NN
Adams Cir 159-J4 H
Adams Cr Dr 58-C1 G
Adams Cr Rd 58-D1 G
Adams Cr View La 58-C1 G
Adams Dr 157-G1, G2 NN
Adams Hunt Dr 72-G7 J
Adams Rd 90-C9 J
Adams St 89-K6; WE-E2 W
Adams Wood La 140-G2 NN
Addingtons 88-C4 J
Addison Rd 159-J3 H
Adelaide St 174-A1 NN
Adele Ct 141-H3 Y
Aden Ct 89-B1 J
Adena Ct 157-K9 NN
Admiral Ct 143-H9 H
Admiral Taussig Blvd 175-H9, J9 NBN
Adrian Cir 159-A6 H
Adriatic Ave 144-F8 H
Adrienne Pl 140-F5 NN
Advocate Ct 124-D8 NN
Adwood Ct 157-K5 H
Agecroft Ct 159-J2 H
Agnes Ct 159-F1 H
Agusta Dr 141-H8 NN
Ainsdale 88-G3 J
Aira Cobra Ct 142-C4 Y
Airborne Dr 142-G10 H
Airborne Rd 158-G1 H
Airport Rd 73-E10 J; 73-E10 Y
Al St 160-C5 H
Alabama La 73-E4 Y
Alamo Ct 143-J10 H
Alan Dr 124-F10 NN
Alaric Dr 144-D10 H
Albacore Dr 125-K3 Y
Albany Dr 158-A5, B5 H
Albemarle Cir 174-B1 NN
Albemarle Dr 104-J3 J
Alberta Dr 140-E3 NN
Alcove Dr 159-H3 H
Alderwood Dr 158-A1 H; 89-B2 W
Alesa Dr 89-C5 J
Alexander Dr 160-C2 H; 140-F2 NN
Alexander Hamilton Blvd 108-H3; YE-C3 Y
Alexander Lee Pkwy 90-E9 Y
Alexander Pl 106-B2 J
Alexander Walker 106-F4 J
Algernourne St 144-C7 H
Algonquian St 92-J9 G
Algonquin Rd 159-B8 H
Alice Ct 125-F4 Y
Alice St 89-G6; WE-C2 W
Alleghany Rd 159-A8, B9 H
Allegheny Rd 72-E8 J
Allen Harris Dr 109-C10 Y
Allen St 159-D8; HE-A8 H
Allendale Dr 160-B1 H
Allendale Pl 90-F5 Y
Allens Mill Rd 125-G2 Y
Allison La 92-G4 G
Allison Rd 140-E3 NN
Allmondsville Rd 58-G8 G
Allyson Dr 73-B8 J
Alma Ct 160-A1 H
Almond Ct 160-B1 H
Almond Dr 157-H2 NN
Alonza Leo Dr 92-G3 G
Alphus St 142-H1 Pq
Alpine St 141-A7 NN
Alta Cres 124-B9 NN
Alton Ct 160-A2 H
Aluminum Ave 158-E10 H
Alva Cir 124-G6 NN
Alvin Dr 160-D1 H
Alwoodley 88-E1 J
Ambassador Dr 157-J3 H
Amber Ct 140-J8 NN; 106-F1 Y
Ambler Ct 159-J2 H
Ambler St 108-J3; YE-C2 Y
Ambrose Hill 106-B2 J
Ambrose Rd 92-G2 G
Ambrose St 160-A5 H
Amelia Ct 141-J4 Y
American Beauty Pl 86-G8 CC
American Legion Dr 124-C8 NN
Americana Dr 141-D8 NN
Amersham Dr 142-E3 Y
Ames Ct 160-B1 H
Ames St 159-G8 H
Ames St E 142-J5 H
Ames St W 142-H5 H
Amesbury La 140-G8 NN
Amherst Ave 157-J7 NN
Amherst Rd 160-A3 H
Amory La 125-F4 Y
Amos Cir 143-E1 Pq
Amy Brooks Dr 140-G10 NN
Amy Ct 160-B3 H
Anacostia Turn 142-C2 Y
Anchor Bay Cove 141-E2 NN
Anchor Dr 126-D2 Y
Anchor La 126-E2 Y
Anchorage Ct 158-A5 H
Anchorage Dr 140-B1 NN
Anchorage La 125-K1 Y
Andalusia Ct 142-E7 H
Anderson Ave 159-A8 H
Anderson Cir 140-J9 NN
Anderson La 125-H3 Y
Anderson Pl 123-F4 F
Andersons Ordinary 106-D5 J
Andover Ct 124-D3 NN; 141-K4 Y
Andover Dr 157-K6 H
Andre Esteve 105-E4 J

Andrew Pl 158-A8 NN
Andrews Blvd 159-H4; 160-A3 H
Andrews Cir 106-A2 J
Andrews Crossing 126-B4 Y
Andrews St 143-F7 L
Andros Isle 142-G7 H
Angel Ct 157-K2 H; 107-D7 J
Angela Ct 160-B1 H
Angelia Way 160-A2 H
Angelo Dr 123-H9 NN
Angus La 143-K9 H
Anita Ct 159-K5 H
Ann Johnson La 87-H10 J
Annapolis Cir 89-F4 Y
Anne Burras La 157-A2 NN
Anne Cir 141-J1 Y
Anne Dr 141-J7 NN
Anne St 160-C5 H
Annette Ct 141-G6 NN
Anns Ct 73-C2 Y
Anthony Dr 159-K5 H
Anthony Wayne Rd 89-B10 J
Antietam Ct 159-G1 H
Antigua Bay 142-G7 H
Antoinette Cir 159-J5 H
Antrim Dr 157-G1 NN
Antrim Rd 91-B6 Y
Apache La 92-H4 G
Apache Tr 143-G10 H
Apollo Dr 159-F2, G1 H
Apothecary Pl 90-F10 Y
Appaloosa Ct 142-D7 H
Appaloosa Dr 142-E4 Y
Appaloosa La 140-F1 NN
Apple Ave 174-H1 H
Apple La 141-G3 Y
Applewood Dr 157-J4 H
April La 157-G1 NN
Aqua Vista Dr 174-E5 NN
Aqueduct Dr 140-H2 NN
Aquia Turn 142-D2 Y
Arabian Cir 142-E4 Y
Arboretum Way 140-J3 NN
Arbutus La 92-H3 G
Arcadia Dr 123-K10; 124-A10 NN
Arch St 159-C7 H; 174-A1 NN
Archer Rd 140-J6 NN
Archers Ct 105-H2 J
Archers Hope Rd 105-G2 J
Archers Mead 106-C3 J
Archibald St 175-J10 NBN
Arden Cir 124-F7 NN
Arden Dr 157-G2 NN; 89-F4 Y
Ardmoor Dr 157-K2 H
Arena St 104-F1 J
Argall Pl 124-H7 NN
Argall Town La 104-A2 J
Argosy Dr 140-B3 NN
Argus St 159-E5; HE-D2 H
Ark Rd 58-G7, H6, K6 G
Arline Dr 123-K7 NN
Arlington Ave 158-B10 NN
Arlington Is Rd 54-G8 J
Arlington Terr 158-B4 H
Armistead Ave 89-G7 W
Armistead Ave N 142-G5, H6, J10; 158-J1; 159-A4; HE-A2 H
Armistead Ave S 159-D7; HE-C3, C4 H
Armor Arch 159-A4 H
Armstead Ave WE-C3 W
Armstrong Ave 159-G9; HE-F7 H
Armstrong Dr 159-C10, D9 H; 106-E1 Y
Arnett Dr 123-K9 NN
Arnold St 158-B9 NN
Arony St 157-K6 H
Arrollton Dr 158-D7 H
Arrow Ct 142-D3 Y
Arrowhead Dr 141-F9 NN
Arrowwood Ct 157-K3 H
Arthur Way 141-E3 NN; 141-E3 Y
Artillery Pl 124-G7 NN
Artillery Rd 159-G1 H; 109-C7 Y
Artists Way 155-B5 I
Ascot 88-H3 J
Ascot Dr 142-E4 Y
Ash Ave 143-B9 L; 174-G2 NN
Ash View 104-A1 J
Ashbury Rd 89-C3 J
Ashford Pl 140-B1 NN
Ashland Ct 140-K8 NN
Ashland La 142-H2 Pq
Ashland St 158-H9 H
Ashleigh Dr 158-G2 H
Ashley Cir 125-K4 Y
Ashley Pl 124-A10; 140-B1 NN
Ashley St 92-H1 G
Ashley Way 88-B10 J
Ashmont Cir 142-B7 H
Ashridge La 140-G4 NN
Ashton Green Blvd 124-A5, B5 NN
Ashton St 90-A8; WE-F4 W
Ashway Cove 140-K5 NN
Ashwood Dr 157-J4 H; 124-G9 NN; 89-B1 W
Aspen Blvd 108-K10; 109-A10; 125-A1 Y
Aspen Ct 105-H2 J
Aspen Dr 124-F5 NN
Aspenwood Dr 158-D5 H
Assembly Ct 156-H1 NN
Aster Way 160-A2 H
Astor Dr 124-A6 NN
Astrid Ct 72-G3 J
Astrid La 72-G3 J
Athens Ave 159-K1 H
Atkins La 141-G5 NN
Atlantic Ave 160-B3 H
Atoka Turn 142-D2 Y
Atwell La 159-F2 H
Auburn La 158-G2 H; 73-F4 Y
August Dr 110-C10 Y
Augusta 88-F5 J
Augusta St 159-B7 H
Augustus Rd 38-G10; 54-G1 NK
Aurel La 72-G1 J
Austin Pt 125-K5 Y
Autozone Way 159-E4 H
Autry Pl 141-A5 NN
Autumn Cir 72-J9 J; 141-A7 NN
Autumn E 72-J10 J
Autumn La 159-J2 H
Autumn Trace 72-J9 J
Autumn W 72-H9 J
Autumn Way 141-J2 Y
Avalon Pier Rd 89-F8 W
Avella Ct 141-K6 NN
Avenue C 124-J8 NN
Averill Ave 159-G8 H
Avery Cres 141-B8 NN
Avis Cir 124-C8 NN
Avon Ave 90-B9 J
Avon Rd 158-E2 H
Avora Ct 124-F6 NN
Aylesbury Dr 124-H4 NN
Aylwin St 175-J8 NBN
Ayrshire Way 124-G9 NN
Azalea Dr 158-J8 H; 140-C3 NN; 109-A9 Y

B

B Ave 124-K8; 125-A9 NN; 109-H6 Y
B C S Dr 141-C5 NN
B St 91-B4 Y
Baccus Ct 160-D1 H
Back Corner Dr 92-K9 G
Back Cr Pk 109-K8 Y
Back Cr Rd 109-J8; 110-A9, B8, B9 Y
Back Landing 143-E2 Pq
Back River La 104-J6 J
Back River Rd 123-C7 F; 159-C5, C7 H
Back River Rd N HE-A5 H
Back River Rd S HE-A6 H
Backspin Ct 141-D2 NN
Bacon Ave 175-K8 NBN; 89-E6; WE-A1 W
Bacon Ct 125-B10; 141-A2 NN
Bacon St 159-D7; HE-B5 H; 108-J4; YE-D4 Y
Baez Ct 123-K10 NN
Bafferton Cir 159-H9 H
Bailey Dr 125-H6 Y
Bailey Rd 125-H7 Y
Bailiff Ct 124-D8 NN
Bainbridge Ave 159-H7 H
Baines Rd 157-K5; 158-A5 H
Baker Blvd 157-J8 NN
Baker Cres 157-J8 NN
Baldric Pl 125-K4 Y
Baldwin Pl 141-A5 NN
Baldwin Terr 158-F6 H
Ballard Dr 107-H6 Y
Ballard La 89-G8; WE-B4 W
Ballard Rd 141-F10 NN; 108-C2 Y
Ballard St 108-J3; YE-C3, D2 Y
Ballou Pl 123-F4 F
Balmoral 88-J5 J
Balmoral Dr 158-K7 H
Balthrope Rd 124-H7 NN
Baltusrol 88-F5 J
Bancroft Dr 159-J4 H
Bandecoot Ct 156-K2 NN
Banister Dr 142-K10; 158-J1 H
Banks La 142-D6 H; 124-A8 NN
Bannaker Dr 90-E10; 106-E1 Y
Bannister Ct 126-H9 Pq
Bannon Ct 158-B6 H
Baptist Rd 108-B4, C5 Y
Baptista Ct 88-F2 J
Barba Dr 159-J1 H
Barbara Ct 124-D4 NN
Barbara La 126-F6 Pq
Barbour Cir 140-K10 NN
Barbour Dr 140-K10 NN
Barcanmore La 110-B5 Y
Barclay Dr 126-C9 Y
Barclay Rd 140-F9, H9 NN
Barcroft Dr 125-G2 Y
Barfleur Pl 73-A8 J; 125-G6 Y
Barfoot Cir 160-C2 H
Barham Blvd 108-G4; YE-A4 Y
Barham Rd 39-E5 NK
Barhamsville Rd 55-K4 J
Barksdale Dr 123-K9 NN
Barksdale Rd 159-D10 H
Barley Mill Pl 88-F6 J
Barlow Rd 73-F3, H7 Y
Barlows Run 73-E1 J
Barn Elm Rd 56-D7 J
Barn Swallow Ridge 125-D4 Y
Barnes Ct 160-C2 H
Barnes La 160-C1 H
Barnes Rd 55-A6, C4, F3 J
Barney Ct 157-F7 NN
Barnstaple Way 73-F3 Y
Baron St HE-F1 H
Barrack St 158-K4 H
Barracks Cir 108-E1 Y
Barracks Rd 108-C2, E1 Y
Barren Pt Rd 58-F6 G
Barrett 106-D1 J
Barretts Ferry Rd 86-J9 CC
Barrie Cir 124-B9 NN
Barrington La 141-E2 Y
Barrington Pl 157-K4 H
Barrister Pl 124-K6 NN
Barron Dr 124-A7 NN; 107-H6 Y
Barron St 159-G5; HE-F2 H
Barrows Mount 104-C2 J
Barry Ct 158-F6 H
Barrymore Ct 158-H2 H
Bartons Landing 155-C5 I
Baseline Rd 107-B9 J
BASF Dr 107-A9, C8 J
Basil Ct 127-B10 Pq
Basil Sawyer Dr 142-H7 H
Bassette St 159-C5; HE-A1, A2 H
Basswood Dr 157-H2 NN
Basta Dr 109-F7 Y
Bastille Ct 89-F6; WE-B2 Y
Bates Ct 88-F1 J
Bates Dr 174-D5 NN
Bates St 158-K9 H
Batson Dr 140-G4 NN
Battery Cir 125-H9 Y
Battery Derussey 160-B9 FM
Battery Dr 92-K10 G
Battery Pl 124-H6 NN
Battle Rd 142-B9, B10 H; 109-B9 Y
Baughman Ct 174-C1 NN
Baxter St 160-A1 H
Bay Ave 158-J10 H
Bay Cliff Ct 141-C1 NN
Bay Club Ct 107-B6 J
Bay E 104-B1 J
Bay Front Pl 160-E1 H
Bay Hill 88-J4 J
Bay Shore La 160-C5 H
Bay St 127-F10 Pq
Bay Tr 140-G1 NN
Bay Tree Beach Rd 110-F9 Y
Bay W 104-B1 J
Bayberry Ct E 159-E2 H
Bayberry Ct W 159-E2 H
Bayberry Dr 141-H6 NN
Bayberry La 89-D9 W; 142-C4 Y
Bayhaven Dr 158-D6 H
Baylor Ct 141-C1 NN
Bayview Dr 126-F6 Pq; 126-B6 Y
Bayview St 160-C4 H
Beach Rd 144-B9, D7; 160-C1 H; 126-H6 Pq; 126-A6 Y
Beach Rd N 109-F6 Y
Beacon Cir 160-C2 H
Beacon Ct 141-F9 NN
Beacon Way 141-E7 NN
Beacons Way 159-G7; HE-G5 H
Beaconsdale La 141-D10; 157-D1 NN
Beakley Pl 123-J6 F
Beall Dr 159-J3 H
Beamer Pl 140-G2 NN
Bear Cr Crossing 159-G2 H
Bears Crossing La 155-B5 I
Beatrice Dr 158-E3 H
Beatties Landing Rd 126-C3 Y
Beaumont St 159-K1 H
Beauregard Hgts 143-H10 H
Beauregard St 90-A8 W
Beaver Castle Ct 157-J2 H
Beaver Rd 91-B8 Y
Beaver Run E 72-G10 J
Beazle La 143-A9 H
Beckie La 107-C7 J
Becouvarakis Ct 143-F10 H
Bedford Ct 160-C1 H
Bedford Rd 141-F10 NN
Beech Dr 54-A7 NK; 157-H3 NN
Beech Tree La 57-F9 J
Beecham Dr 125-H9 Y
Beechmont Dr 159-H3 H; 123-J10; 124-B9 NN
Beechnut Ct 104-A1 J
Beechwood Ave 174-G3 NN
Beechwood Dr 54-G10 J; 89-B2 W; 125-H7 Y
Beechwood Dr W 107-D6 J
Beechwood Hills 123-J9 NN
Beechwood La 142-C4 Y
Beechwood Rd 158-E5 H
Beechwood Rd E 107-H5 Y
Beler Rd 88-K8 J
Belfast Ave 158-J9 H
Belgrave Rd 140-D3 NN
Belinda Dr 157-E2 NN
Bell Hill Dr 125-E2 Y
Bell King Rd 141-A5 NN
Bell St 158-K8; 159-A8 H
Bellamy Pl 124-G6 NN
Belle Cir 124-A9 NN
Belle La 143-F1 Pq
Belle Meade Ct 140-D5 NN
Bellehaven Dr 92-J9, K9 G
Belleview 88-K5 J
Bellfield Dr 124-C4 NN
Bellfield Rd 91-H7, J8; 92-A8 Y
Bellgate Ct 141-B1 NN
Bellgrade Dr 142-D7 H
Bellis Cove Dr 126-G10 Pq
Bellows Pl 142-A4 Y
Bellows Way 141-A1, A2 NN
Bells Is Dr 144-D5, D6 H
Bellview Terr 159-F3 H
Bellwood Rd 157-J5 H; 157-H6 NN
Belmont Cir 141-J4 Y
Belmont Dr 55-D2 J
Belmont Pl 158-A5 H
Belmont Rd 157-F3 NN
Belray Dr 141-H8 NN
Belton Pl 124-H5 NN
Belvedere Dr 157-J9 NN
Belvin La 110-C5 Y
Belvoir Cir 124-H5 NN
Bending Oak Dr 158-D5 H
Benedict Ave 143-E9 L
Benjamin Ct 73-A10 J
Benjamin Terr 158-D4 H
Bennett Cir 106-B4 J
Bennett Farm Rd 126-H7 Pq
Bennett Rd 127-A9 Pq
Bennetts Pond Rd 103-J1 J
Bennington Ct 125-D10 Y
Benns Rd 141-G8 NN
Benson Dr 144-D10 H
Bent Branch La 124-A8 NN
Bent Cr Rd 104-B1 J
Benthall Rd 160-D3 H
Bentley Dr 158-J1 H; 140-B1 NN
Bently Ct 142-F4 Y
Benton La 159-A6 H
Bergen Cir 72-G2 J
Berger Pl 141-C8 NN
Berkeley La 89-C9 W
Berkeley Pl 124-H7 NN
Berkeley Town Rd 55-B8 J
Berkeleys Green 104-B1 J
Berkely Cir 104-B3 J
Berkley Dr 160-A5 H; 90-E8 Y
Berkshire Dr 141-B1 NN
Berkshire Rd 88-J1 J
Berkshire Terr 158-C3 H
Bermuda Cir 104-B2 J
Bernard Ave 159-D4 H
Bernard Dr 140-G3 NN
Bernard Rd 160-A10 FM
Bernardine Dr 124-H8 NN
Berrets Pt Rd 103-B1 J
Berrow 88-G3 J
Berrys Landing 126-B5 Y
Bertha Ct 142-E7 H
Berwick N 88-F2 J
Berwick St 142-J1 Pq
Bessies Landing Dr 126-G6 Pq
Bethany Ct 124-B10 NN
Bethany Terr 142-A1 Y
Bethel Ave 159-D4 H
Bethel Rd 124-A4 NN
Bethune Dr 140-K7 NN; 90-E10 Y
Betty Ct 158-G6 H
Betty Lee Pl 124-K6 NN
Betz La 142-A6 H
Betzer Dr 107-G6 NN
Beverly Hills Dr 140-F10 NN
Beverly St 159-E1 H
Bibb Rd 109-D6 Y
Bickerton Ct 124-G4 NN
Bickfield Dr 159-A2 H
Bickford St 159-H8 H
Big Ben Ct 124-H5 NN
Big Bethel Pl 142-B10 H
Big Bethel Rd 142-B9; 158-B4, C5 H; 142-A1 Y
Big Gap Rd 73-H3 Y
Big Oak La 58-B1 G
Biggins Cir 157-H8 NN
Billeves Rd 141-A6 NN
Biltmore Ct 140-K4 NN
Bimini Crossing 142-G7 H
Binnacle Dr 140-B2 NN
Birch Ave 158-G10 H
Birch Cir 56-H8 J
Birch Dr 157-H3 NN
Birch St 92-H2 G; 143-B8 L
Birchwood Ct 124-C3 NN
Bird La 157-F2 NN
Birdella Dr 72-G7 J; 158-B9 NN
Birdie 88-E3 J
Birdie La 141-D2 NN
Birkdale 88-H4 J
Birkdale Ct 141-D1 Y
Birmingham 88-H3 J
Bishop Ct 125-C10 NN
Black Ave 159-G8 H
Black Horse La 38-B8 NK
Black Oak Ct 158-A3 H
Black Oak Dr 105-G1 J
Black Twig Dr 140-D3 NN
Black Widow Dr 142-C7 Y
Blackberry Bend 141-J4 Y
Blackberry La 143-K9 H
Blackburn La 142-B8 H
Blackfoot Ct 87-J8 J
Blackheath 88-B2, B3 J
Blacklake 88-H5 J
Blackmore Pl 158-K3 H
Blacksmith Arch 142-A4 Y
Blacksmythe La 140-D6 NN
Blackwater La 143-F9 H
Blackwater Way 141-C8 NN
Blackwood Ct 142-E5 Y
Blair Ave 174-H2 H; 174-G3 NN
Blair Ct 106-F4 J
Blair Dr 109-C9 Y
Blair St WE-E3 W
Blake Cir 159-H2 H
Blake Loop 141-C8 NN
Blalock Dr 90-E8 Y
Bland Blvd 124-H10, K9; 140-G1 NN
Bland St 159-D3 H
Blankenship La 157-J4 H
Blanton Dr 123-K4 NN; 110-B10 Y
Blarney La 107-H10 NN
Blassingham 106-A2 J
Blazer Ct 158-F6 H; 124-J6 NN
Blenheim 88-H4 J
Blenheim Ct 124-G4 NN
Blount Ct 159-J6 H
Blount Pt Rd 140-G10 NN
Blow Flats Rd 107-D8 J
Bloxom Dr 123-J9 NN
Bloxoms La 160-C1 H
Blue Bill Run 72-G10 J
Blue Crab Dr 126-J8 Pq
Blue Crab Rd 141-D6 NN
Blue Heron Dr 126-A3 Y
Blue Heron Tr 140-G6 NN
Blue Marlin Way 160-E2 H
Blue Pt Terr 141-A4 NN
Blue Ridge Ct 72-E9 J
Blue Ridge Rd 158-D3 H
Blue Sage Path 160-A2 H
Blue Tulip La 144-B7 H
Blunt Ct 140-F10 NN
Bob Gray Cir 158-D2 H
Bobcat Dr 142-C5 Y
Bobs Ct 157-K4 H
Bobwhite Ct 92-G1 G
Boca Raton Cir 106-G1 Y
Bodiche Ave 107-A10 J
Boeing Ave 159-D1 H
Bogey Dr 88-E3 J
Bohnert Dr 159-B1 H
Bolden Dr 92-J6 G
Bolivar Dr 109-D9 Y
Bolling Blvd 155-K10 I
Bolling Rd 90-D9 J
Bollman Rd 108-A1, B2 Y
Bonaire Dr 159-J1 H
Bond Cir 140-E3 NN
Bond St 158-C2 H
Bonifay Dr 158-A5 H
Bonita Dr 144-F8 H; 140-D4 NN
Bonito Dr 125-K3 Y
Bonneville Dr 144-C10 H
Bonney Way 159-K2 H
Bonnie La 140-J7 NN
Bonnie Lee Pl 157-K8 NN
Bonwood Rd 158-E3 H
Bonyman Ct 88-H1 J
Booker St 159-J7 H
Booker-T Rd 90-E10 Y
Boone Pl 157-K10 NN
Booth Cir 141-A10 NN
Booth Rd 141-A9 NN
Bosch La 141-B8 NN
Boston Common 89-F4 Y
Bosun Ct 140-C2 NN
Boswell Dr 159-D3 H
Botetourt Rd 157-E4 NN
Botetourt St 89-J7; WE-E3 W
Boucher Dr 107-H6 Y
Boulder Dr 124-A5 NN
Boulder Way 90-C6 Y
Boundary Rd 107-K5 NN; 107-K5; 108-A4, D5 Y
Boundary Rd W 107-A4, E6 J; 107-E6 NN; 106-J2; 107-E6 Y
Boundary St N 89-G7; WE-C3 W
Boundary St S 89-G7, G8; WE-C4 W
Bounty Cir 144-B10 H
Bowen Dr 158-C8 H
Bowen St 143-E8 L
Bowie Ct 124-A8 NN
Bowman Terr N 141-K1 Y
Bowman Terr S 141-K1 Y
Bowstring Dr 90-D5 Y
Box St 158-J10 H
Boxelder Ct 158-A3 H
Boxley Blvd 140-G4 NN
Boxwood La 88-F5 J; 140-F2 NN
Boxwood Pt Rd 159-G4 H
Boxwood St 159-G6 H
Boxwood St N HE-G2 H
Boxwood St S HE-G3 H
Boyd Cir 124-C10 NN
Boykin La 160-B2 H; 124-H10; 140-H1 NN
Bozarth Ct 89-K6; WE-E2 W
Bracken Rd 108-E1 Y
Brackin Ct 159-J3 H
Brackin La 124-B7 NN
Brad Ct 124-A10 NN
Braddock Ct 89-A10 J
Braddock Rd 174-K1 H; 89-A10; 105-A1 J
Bradford Cir 140-J4 NN
Bradley Dr 110-C7 Y
Bradley La 160-A4 H
Bradmere Loop 123-J4 NN
Bradshaw Ct 143-C1 Pq
Bradshaw Dr 88-K7; 89-A7 J
Brady Dr 88-A5 J
Braeburn La 92-D1 G
Braemar Dr 158-K7 H
Bragg Ct 158-D7 H
Bramston Dr 159-A2 H
Brancaster 88-H3 J
Branch La 109-E9 Y
Branch Rd 92-F1 G
Branch Rd E 109-D7 Y
Branchs Pond Rd 71-K1 J
Brandon Cir 104-K2 J
Brandon Ct 159-H2 H
Brandon Rd 157-F5 NN
Brandon Way 125-H8 Y
Brandsby Ct 124-G4 NN
Brandywine Dr 141-B3 NN; 125-H4 Y
Brannon La 88-K7 J
Branscome Blvd 104-J4 J
Bransford Ct 88-H2 J
Brantley Ct 141-K5 Y
Brantleys Trace 157-J4 H
Brassie Dr 125-E10; 141-E1 Y
Brassie Way 125-C10 NN
Braxton Ct 89-G7; WE-B3 W; 142-A5 Y
Braxton Rd 40-D9 NK
Bray Rd 92-H6 G
Bray Wood Rd 106-D3 J
Breakwater St 159-E8; HE-C8 H
Breckinridge Ct 158-D6 H
Breezy Pt 126-K7 Pq
Breezy Pt Dr 125-K8 Y
Breezy Tree Ct 124-A6 NN
Brenda Rd 141-G9 NN
Brennans La 107-A6 J
Brennhaven Dr 140-D5 NN
Brentmeade Dr 142-D2, E3 Y
Brentmoor Ct 124-H5 NN
Brentwood Dr 159-E4 H; 157-H3 NN
Bret Harte Dr 140-F3 NN
Brettwood Ct 90-B9 W
Brewer St 157-J7 NN
Brewhouse Ave 89-F10 W
Brian Ct 107-C7 J
Brian St 88-A6 J
Brians Ct 144-C9 H
Briar Dr 158-G7 H
Briar La 72-H6 J
Briar Patch Pl 157-A1 NN
Briarfield Rd 158-F7 H; 158-A9 NN
Briarwood Ave 107-C6 J
Briarwood Dr 158-D7 H
Briarwood Pl 125-A1 Y
Brick Bat Rd 87-G9 J
Brick Ch Rd 125-D3 Y
Brick Cr La 142-F6 H
Brick Kiln Blvd 140-K1; 141-A2 NN
Brickhouse La 142-H2 Pq
Brickyard Rd 54-K10; 70-K1 J
Bridge Crossing 109-A10 Y
Bridge Ct 86-J9 CC
Bridge La 109-A10 Y
Bridge St 159-E7; HE-D8 H
Bridge Wood Dr 141-H4 Y
Bridgeport Cove Dr 160-A2 H
Bridgewater Dr 89-A8 J; 107-H10 NN
Bridle La 124-E2 NN
Bridle Path 58-J4 G
Brigade Dr 125-J4 Y
Brigham St 159-J5 H
Bright St 159-B8 H; 89-F7; WE-B3 W
Brighton Cir 140-D4 NN; 109-K9 Y
Brighton La 141-G5 NN
Brightwood Ave 159-B8 H
Brigstock Cir 141-E7 NN
Brinkman Dr 142-K10 H
Bristol Cir 88-J10, K10; 104-J1 J
Bristol Ct 158-B3 H
Bristol La E 142-E4 Y
Bristol La W 142-D4 Y
Britnie Ct 140-G2 NN
Brittain La 143-J7 H
Brittania Dr 90-D7 Y
Brittington E 104-C1 J
Brittington W 104-B1 J
Broad Bay Cove 141-D2 NN
Broadleaf Dr 89-H10 W
Broadmead Ct 89-D5 J
Broadmoor 88-G5 J
Broadstreet Rd 158-J4; 159-A4 H
Broadwater 88-J4 J
Brockridge Hunt Dr 158-D2 H
Brockton Ct 89-D10 W
Brogden La 158-A2 H
Brokenbridge Rd 109-A10 Y
Bromley Dr 159-H1 H; 104-J2 J
Brompton Ct 124-G5 NN
Bromsgrove Dr 158-D2 H
Bronco Dr 142-C7 Y
Brook La 125-J7 Y
Brook Rd 109-A10 Y
Brook Run 125-H10 Y
Brooke Ct 88-G1 J
Brooke Dr 159-E8 H; 107-H6 Y
Brooke St 157-J7 NN
Brookfield Dr 159-A3 H; 140-D4 NN
Brookhaven Dr 89-A7 J
Brookman Pl 144-D8 H
Brooks St 89-E6, E7; WE-A2, A3 W
Brookside Dr 141-A2 NN
Brookstone Ct 141-C1 Y
Brookwood Dr 89-D10; 105-C1 J; 89-D10 W
Broomfield Cir 106-B4 J
Brough La 159-G6; HE-G3 H
Broughton La 143-D10 H
Brout Dr 158-A5 H
Brower Rd 89-B4 J
Brown Cir 160-A3 H
Browns Dr 71-K3 J
Browns Family La 92-G2 G
Browns La 108-A4 Y
Browns Marina Rd 155-K10 I
Browns Neck Rd 126-H8 Pq
Brucker Rd 54-B3 NK
Brunell Dr 158-A5 H
Brunk Dr 89-C4 J
Brunswick Pl 141-J7 NN
Bruton Ave 141-G8 NN
Bruton Dr 90-C7 Y
Bryan Ct 141-A7 NN
Bryant Dr 159-K6 H
Bryant St 143-D8 L
Bryson Ct 124-E6 NN
BSM Rd 107-B10 J
Buchanan Dr 159-D9 H; 124-F8, H7 NN
Buchanan St 175-K8 NBN
Buck Cir 124-F4 NN
Buckeye La 92-G1 G
Buckingham Dr 158-J7 H; 110-C7 Y
Buckingham Green 141-B2 NN
Buckner St 108-J3; YE-D2 Y
Buckroe Ave 160-C3 H
Bucktail Run 125-E4 Y
Bucktrout La 89-J7; WE-E4 W
Budweiser St 158-H8 H
Buffalo Dr 144-C10; 160-C1 H
Buffie Rd 125-H9 Y
Bufflehead Cove 144-A8 H
Bugle Ct 141-F1 Y
Build America Dr 158-G4 H
Bulkeley Pl 157-G5 NN
Bull La 157-A1 NN
Bullard St 123-H6 F
Bullock Pl 141-K8 H
Bunche Dr 90-E10 Y
Bundy St 123-F6 F
Bunker Arch 88-D10 J
Bunker Hill 175-H8 NBN
Bunker Hill Cir 141-C4 NN
Bunker St 160-B9 FM
Bunting La 127-C10; 143-C1 Pq
Bunting Pt Rd 126-C8 Y
Burbank St 89-G6; WE-C2 W
Burcher Rd 140-G8 NN; 125-H4 Y
Burgess Ave 160-B4 H
Burgess St 90-C9 J
Burgh Course 88-F3 J
Burgh Westra Dr 159-H2 H
Burghley Ct 124-H5 NN
Burke Ave 141-D10 NN
Burma Rd 90-J10; 91-A8, D7; 106-J2 Y
Burman Wood 158-A1 H

Burnette Dr 160-B3 H
Burnham 88-G3 J
Burnham Pl 141-B9 NN
Burnham Rd 90-A7 Y
Burnley Dr 89-A8 J
Burns Ave 141-G8; 157-G6 NN
Burns La 89-F8; WE-A4 W
Burns St 159-D9 H
Burnt Bridge Way 109-A10 Y
Burnt Run 109-A10 Y
Burnwell Ct 106-B3 J
Burnwether La 105-D4 J
Burrell Loop 143-C10 L
Burrell St 143-B10, C9 L
Burrows Ct 104-F3 J
Burtcher Ct 106-E4 J
Burton Ct 72-D10 J
Burton St 158-B2 H
Burton Woods Dr 72-D10 J
Burts Rd 125-F7 Y
Burwell Cir 141-A10 NN
Burwick Ct 159-K1 H
Busch Gdns Blvd 106-G3 J
Bush Neck Rd 87-D5 J
Bush Rd E 142-J4 H
Bush Rd W 142-H4 H
Bush Springs Rd 56-C10; 72-C1 J
Bush St 158-K4; 159-A4 H
Butler 88-F6 J
Butler Dr 158-D7 H
Butler Farm Rd 142-E10, G10; 158-D1 H
Butler Pl 157-B2 NN
Butner St 123-J7 F
Butte Cir 124-B9 NN
Buttercup La 140-H3 NN
Butternut Dr 158-C5, D5 H
Buttonwood Dr 158-D4 H
Buttonwood La 142-C5 Y
Buxton Ave 174-F1, G2 NN
Bypass Rd 89-F5; WE-E1 W; 89-F5; 108-C2 Y
Byrd La 88-H8 J; 141-G2 Y
Byrd St 158-F10 H
Byron Rd 108-K10 Y

C

C Ave 109-H6; 110-A6 Y
C St 91-C4 Y
Cabarrus Ct 141-D8 NN
Cabell Ct 72-J10 J
Cabell Dr 140-D5, D6 NN
Cabell La 144-B8 H
Cable Way 142-C5 Y
Cabot Dr 159-J1 H; 108-C4 Y
Cades Ct 140-J5 NN
Caffee Cr La 58-F5 G
Caisson Crossing 159-G1 H
Caldroney Dr 140-G3 NN
Caldwell Dr 158-G1 H
Cale Cir 157-B1 NN
Calhoun St 159-H5 H
Calla Ct 124-E6 NN
Callahan Dr 106-F1 Y
Callahan Walk 157-E3 NN
Callis La 126-E8 Pq
Calm Branch Ct 141-C1 NN
Calthrop Neck Rd 126-C9, C10; 142-B10 Y
Calthrop Pt 126-C7 Y
Calumet Turn 142-C2 Y
Calvary Terr 158-F5 H
Calvert Dr 141-G8 NN
Calvert St 159-F1 H
Calvin St 89-C5 W
Camberley Cir 89-A8 J
Cambertree Way 124-E5 NN
Cambridge Cres 159-C10 H
Cambridge Ct 140-D1 NN
Cambridge La 90-D5 Y
Cambridge Pl 159-G5 H
Camden Cir 88-B10 J
Camden St 159-E8 H
Camden Way 142-E3 Y
Camellia Dr 140-F6 NN
Camelot Cres 125-H10 Y
Camelot Ct 141-F4 NN
Cameo Dr 158-H2 H
Cameron Dr 140-J9 NN
Cameron Pl 123-F4 F
Cameron St 159-H7 H
Camille Ct 141-K5 Y
Camp Okee Dr 92-J10, K10 G
Camp Rd 55-C5 J
Campbell Close 105-D4 J
Campbell La 124-G10; 140-F1 NN
Campbell Rd 140-G1 NN
Campside La 108-A8 NN
Campsite Dr 108-B10 NN
Campton Pl 124-H5 NN
Campus Dr 89-E7, F8; WE-A2, A3, A4 W
Camrose Dr 104-B1 J
Canal Dr 143-B1 Pq
Canal Rd 144-D7 H
Canal St 70-G2 J
Canavan Dr 159-J3 H
Canberra Ct 142-C7 Y
Candle La 124-C6 NN
Candlestick Pl 106-F1 Y
Candlewood Dr 158-B6, C6 H
Candlewood Way 141-D8 NN
Canford Dr 159-H5 H
Canham Rd 105-B1 J
Cannister Ct 159-G1 H
Cannon Dr 140-D5 NN
Cannon Rd 109-A8 Y
Cannonball Ct 90-B6 Y
Canoe La 92-K10 G
Canon Blvd 141-D4, E6 NN
Cantamar Ct 160-D4 H
Canter Cir 140-D6 NN
Canterbury La 89-D8 W
Canterbury Pl 73-B10 J
Canterbury Rd 158-J4; 159-A4 H
Canterbury Run 140-D1 NN
Canton Dr 159-J5 H
Canvas Back Run 72-G10 J
Canvasback Tr 140-F6 NN
Cape Dory Dr 158-F2 H
Capitol Ct 89-K7; WE-F3 W
Capitol Landing Rd 89-K6, K7; 90-A4 W; 90-A4 Y
Capitol Ldg Rd WE-E3, F1 W
Cappahosic Rd 58-J10 G
Capps Quarter 159-E8 H
Captain Cleve La 92-G3 G
Captain Drew 88-B7 J
Captain Graves 106-F4 J
Captain Jack La 144-B6 H
Captain John Smith Rd 104-J5 J; 156-K2; 157-A2 NN
Captain Newport Cir 89-C2 W
Captain Wynne Dr 104-C3 J
Captains Ct 143-H9 H; 89-G9 W
Captains La 140-C4 NN
Caran Rd 89-F4 Y
Cardiff St 158-B2 H
Cardinal Acres Dr 104-F3 J
Cardinal Ct 103-K2 J
Cardinal Dr 144-C10 H
Cardinal La 141-G10 NN; 125-G10 Y
Caribou Dr 141-E8 NN
Carla Dr 123-H9 NN
Carleton Rd 107-H9 NN
Carleton St 90-A8 W
Carlisle Ct 157-H3 H
Carlisle Mews 89-B3 J
Carlton Ct 89-C3 W
Carlton Dr 158-B5 H; 126-A8 Y
Carmel Terr 157-K4 H
Carmel Valley 88-D2 J
Carmella Cir 124-D10 NN
Carmen Dr 160-C2 H
Carmine Pl 142-J10 H
Carmines Ct 126-K8 Pq
Carmines Is La 92-E5 G
Carmines Is Rd 92-E5 G
Carnation Dr 124-D10 NN
Carnegie Dr 141-B7 NN
Carnegie St 159-H8 H
Carnoustie 88-F3 J
Carnoustie Ct 141-D1 Y
Carol Dr 125-E2, E3 Y
Carol La 107-C7 J
Carolina Ave 159-B6 H; 175-J9 NBN
Carolina Blvd 89-A9 J
Carolyn Cres 125-H2 Y
Carolyn Dr 141-B9 NN
Carolyn La 92-J4 G
Carpenter Dr 126-D9 Y
Carr Dr 158-B4 H
Carraway Terr 125-G6 Y
Carriage Dr 160-C2 H
Carriage Hill Dr 126-E10 Pq
Carriage Hill Dr N 126-E10 Pq
Carriage La 141-D8 NN
Carriage Rd 89-C5 J
Carrington Ct 157-J2 H
Carroll Dr 126-F6 Pq
Carroll St 159-J5 H
Carrs Hill Rd 89-F4 Y
Carson Cir 140-H5 NN
Carter Rd 54-F4 NK
Carters Grove Country Rd 105-J1; 106-C1, H5; 107-A6 J; 89-J8 W
Carters Grove Ct 159-H9 H
Carters Neck Rd 73-H5 Y
Cartgate 88-K3 J
Carver Dr 157-J5 NN
Carver Pl 125-E2 Y
Carver St 159-D5; HE-B1 H
Cary St 89-F8; WE-B4 W
Carys Chapel Rd 142-E2 Y
Carys Trace 141-J1 Y
Carywood La 140-B1 NN
Cascade Dr 124-A9 NN
Cascade View Ct 157-H2 H
Cascades 88-F4 J
Casey Marie La 157-F7 NN
Casey Terr 157-H1 NN
Casper La 141-D2 NN
Casseday Way 124-C4 NN
Castel Pines 88-H5 J
Castellow Ct 125-J6 Y
Castle Dr 124-E5 NN
Castle Haven Rd 157-J2 H
Castle Keep Ct 124-G6 NN
Castle La 104-F1 J
Castle Rd 159-E4 H
Castlewood Ct 159-H2 H; 125-J3 Y
Catabaw Ct 90-D8 Y
Catalina Dr 144-C10 H; 124-A9 NN
Catalpa Ave 158-H10; 174-H1 H
Catalpa Dr 73-B8 J; 157-H3 NN
Catesby Jones Dr 159-D9 H
Catesby La 90-A7, B8 Y
Catherine Ct 125-J1 Y
Cathy Dr 123-K6 NN
Catina Way 124-F9 NN
Cattail La 142-E6 Y
Cavalier Dr 124-F5 NN; 125-F7 Y
Cavalier Rd 159-E3 H
Cay St 157-K6 H
CC Spaulding Dr 158-G5 H
CEBAF Blvd 141-C5 NN
Cedar Ave 143-B9 L; 174-G2 NN
Cedar Bush Rd 92-A1 G
Cedar Cir 72-H6 J
Cedar Cr La 88-G9 J
Cedar Ct 105-H2 J
Cedar Ct N 124-C5 NN
Cedar Ct S 124-C5 NN
Cedar Dr 159-F2, F3 H; 54-H10 J
Cedar Glen Ct 141-A2 NN
Cedar La 157-D4 NN
Cedar Pt Cres 126-B7 Y
Cedar Pt Dr 159-E8; HE-D8 H
Cedar Pt La 56-H8 J
Cedar Rd 127-A10; 143-A1 Pq
Cedar Rock 88-G5 J
Cedar Run 104-B1 J
Cedarwood Dr 92-J5 G
Cedarwood La 88-E6 J
Cedarwood Way 124-H7 NN
Celey St 158-K8 H
Cellardoor Ct 158-C7 H
Cemetary La 90-F9 Y
Cemetery La 158-J10 H
Cemetery La W 143-E3 Pq
Cemetery Rd 159-F7; HE-F7 H
Center Ave 157-H7 NN
Center La 89-F8 W
Center St 159-F6; HE-E3 H; 141-B8 NN
Centerville Rd 72-D10, F7; 88-A7, B2, B5 J
Central Pkwy 157-A3 NN
Central St 160-A6 H
Cesare Ct 123-J9 NN
Cessna Cir 124-K9 NN
Chad La 157-K9 NN
Chadds Cir 125-K4 Y
Chadwick Ct 141-K5 Y
Chadwick Dr 159-A2 H
Chadwick Pl 141-B9 NN
Chalice Ct N 124-E5 NN
Chalke Ct S 124-E5 NN
Challenger Way 142-G4 H
Chalmer Ct 158-D7 H
Chamberlin Ave 159-J6 H
Champions Path 142-E2 Y
Chancellor Rd 158-J10 H
Chanco Dr 140-J7 NN
Chanco Rd 89-A10 J
Chandler Ct 89-G7; WE-B4 W
Chandler La 108-B5 Y
Chandler Pl 140-B1 NN
Channel Dr 141-D7 NN
Channel House Ct 107-B6 J
Channel La 160-E2 H
Channelwalk Dr 143-C1 Pq
Chanterain Close 89-G9 W
Chanticlair Dr 141-J4 Y
Chanticlar Ct 124-E6 NN
Chanticleer Ct 90-A6 W
Chantideer Ct WE-F2 W
Chantilly Ct 143-H10 H
Chapel Hill La 73-F4 Y
Chapel St 159-C7; HE-A6 H
Chapin Wood Dr 124-C3 NN
Chapman Dr 92-H5 G
Chapman Rd 90-G6, H7 Y
Chapman Way 124-E4 NN
Chaptico Run 142-D2 Y
Charing Cross 89-B3 J
Charitable 88-F4 J
Charity La 140-H3 NN
Charlemagne Ct 124-E5 NN
Charlene Ct 125-F6 Y
Charlene Loop 158-D2 H
Charles Cir 110-D10 Y
Charles Dillard La 89-D3 W
Charles Parish Dr 126-G9 Pq
Charles Rd 125-J6 Y
Charles River Landing Rd 90-F4 Y
Charles St 159-D8 H; 123-J4, K3 NN
Charleston Pl 104-E1 J
Charleston Way 140-J5 NN
Charlotte Dr 141-H6, H7 NN
Charlton Dr 159-A1, A3 H
Charolais Run 143-K9 H
Charter Cir 141-D8 NN
Charter Dr 126-B5 Y
Charter House La 88-F7 J
Charter Oak Dr 124-E7 NN
Chartwell Dr 124-H6 NN
Chase Ct 124-B10 NN
Chase N 104-B1 J
Chase Oak Ct 141-E2 Y
Chase Rd 90-H5, J4; 91-A3, C3 Y
Chase S 104-B1 J
Chateau Dr 73-C3 Y
Chatham Ct 125-H3 Y
Chatham Dr 124-H10; 140-H1 NN
Chatham Terr 158-F5 H
Chatsworth Dr 141-H7, J6, J8 NN
Chattanoga Ct 143-H10 H
Cheadle La 144-B9 H
Cheadle Loop Rd 110-A10; 126-A1, B1 Y
Cheadle Pt Rd 126-B1 Y
Cheese Spring Rd 156-K2 NN
Cheesecake Rd 107-A3 J; 106-K3; 107-A3 Y
Cheeseman Rd 90-E9 Y
Chelmsford Way 140-H5 NN
Chelsea Cres 72-K7 J
Chelsea Pl 107-J9 NN
Chelsea Rd 89-G5 Y
Chelsford Way 88-B10 J
Cheltenham Way 142-E3 Y
Cherbourg Dr 140-J7 NN
Cherokee Ct 124-A9 NN
Cherokee Dr 124-J9; 125-A8 NN; 126-E9 Pq
Cherokee Rd 159-B9 H
Cherokee Tr 92-H4 G
Cherry Acres Dr 159-E5; HE-D1 H
Cherry Ave 174-H1 H
Cherry Cr Dr 124-D3 NN
Cherry Pt Dr 126-B7 Y
Cherry Row La 41-F8, G5 KQ
Cherry St 143-C8 L
Cherrywood Ct 124-G5 NN; 89-B2 W
Cherwell Ct 73-D2, E3 Y
Cheryl Cir 140-F3 NN; 126-J8 Pq
Chesapeake Ave 159-A10, C10, D9 H; 56-A10 J; 174-G3 NN
Cheshire Ct 140-C3 NN
Cheshire La 157-J3 H
Chessie St 157-J8 NN
Chester Rd 124-D3 NN
Chesterfield Rd 158-K10; 174-K1 H
Chestnut Ave 158-B6 H; 143-B8 L; 158-B6, C9; 174-D1 NN
Chestnut Ct 125-H6 Y
Chestnut Dr 105-C1 J
Chestnut Hill Ct 88-A10 J
Cheyenne Dr 124-F4 NN
Chichester Ave 159-G5 H
Chickahominy Bluff Rd 86-K9 CC
Chickahominy Ct 54-F8 NK
Chickahominy Dr 56-B10 J
Chickahominy Rd 71-J4; 72-A3 J
Chickamauga Pike 159-G2 H
Childress Dr 140-B1 NN
Childs Ave 158-G9 H
Chinaberry Pl 158-J1 H
Chinaberry Way 125-E5 Y
Chincoteague Dr 158-K10; 159-A10 H
Chinkapin La 56-G7 J
Chinkapin Tr 124-E6 NN
Chinook Ct 124-A8 NN
Chinquapin Orchard 126-D10 Y
Chipanbeth Ct 143-K10 H
Chipley Dr 124-F10 NN
Chippendale Ct 158-B6 H
Chippenham Dr 142-D4 Y
Chipper La 144-C9 H
Chippokes Turn 142-B3 Y
Chischiak Watch 108-J3; YE-D2 Y
Chisel Run Rd 73-C10; 89-B1 J
Chiskiake St 92-K8 G
Chisman Cir 126-G2 Y
Chisman Dr 160-A2 H
Chisman Landing Rd 125-K1 Y
Chisman Pt Rd 126-C1 Y
Chisom Cir 124-E2 NN
Chiswick Cir 124-J5 NN
Chiswick Pk 88-H1 J
Choice La 141-D2 NN
Choisy Cres 125-G6 Y
Choppers Rd 103-D1 J
Choptank La 140-J2 NN
Choptank Turn 142-B2 Y
Chowan Cir 126-F9 Pq
Chowan Pl 124-F5 NN
Chowan Turn 142-C2 Y
Chowning Dr 160-C2, D2 H
Chowning Pl 125-H10 Y
Chris Slade Chase 142-F2 Y
Christian St 124-B6 NN
Christin Way 141-J8 H
Christine Cir 124-B9 NN
Christine Ct 158-F2 H
Christopher Newport Ave 92-K9 G
Christopher Newport Dr 157-E3 NN
Christopher Pl 174-E5 NN
Christopher Wren Rd 89-C2 W
Chuckatuck Run 142-C2 Y
Church La 144-B8 H; 56-A10; 71-J2, K1 J
Church Rd 140-J7 NN; 108-C4 Y
Church St 106-K5; 107-A5 J; 143-C1 Pq; 108-J3; YE-D3 Y
Churchill La 124-H6 NN
Churchill Terr 158-B2 H
Cicada Ct 158-B6 H
Cindy Cir 140-G3 NN
Cindy Ct 157-K2 H
Circle Dr 157-J8 NN
Circuit La 124-E8 NN
Cisco Ct 159-F1 H
Citation Dr 124-K9 NN
City Farm Rd 140-E7 NN
City Hall Ave 142-G1 Pq
City Line Rd 158-D9 H; 158-D9 NN
Civil Ct 124-E8 NN
Claiborne Pl 140-K5 NN
Claire La 140-K3 NN
Clara Croker 106-B2 J
Clarden Ct 125-G6 Y
Claremont 105-K3 J
Claremont Ave 174-J1 H
Clarendon Ct 88-J1 J
Clark La 73-B8 J; 109-K10 Y
Clark Rd 144-B9 H
Clarke Ave 143-C5 L
Clarke Ct 88-G2 J
Claudia Dr 92-G4 G
Claxton Cr Rd 110-C9 Y
Claxton Terr 144-D10 H
Clay Cir 88-A6 J
Clay Dr 141-A4 NN
Clay St 159-J7 H
Claymill Dr 141-B2 NN
Claymore Dr 159-D2 H
Clayton Dr 159-E4 H; 141-E1 Y
Clearbrook Rd 124-C4 NN
Clearbrooke Landing 125-E5 Y
Clearwater Ct 125-C10 NN
Clements La 125-J4 Y
Clemson Dr 124-B9 NN
Clemwood Pkwy 159-F1 H
Cleopatra Ct 159-G1 H
Cliffe Ct 54-G6 NK
Clifton Ave 92-H1 G
Clifton Ct 124-B10 NN
Clifton St 158-J10; 174-J1 H
Cline Dr 157-K4 H
Clinton Cir 158-G1 H
Clinton Dr 158-A9 NN
Clipper Dr 143-J9 H; 140-C2 NN
Clovelly La 159-C7 H
Clover St 159-E1 H
Cloverdale Rd 107-H4 Y
Cloverleaf La 56-J10 J; 141-H7 NN
Club Run Blvd 159-J2 H
Club Terr 156-K1; 157-A1 NN
Club Way 110-A9 Y
Clubhouse Way 124-E2 NN
Cluster Pl 88-H7 J
Clyde St 159-D9 H
Clydesdale Ct 142-D7 H
Clydesdale Dr 142-E3 Y
Clydeside 88-K3 J
Coach House Dr 89-A2 J
Coach Hovis Dr 142-F2 Y
Coach St 160-D2 H
Coach Tr 124-J6 NN
Coachman Cir 123-J10 NN
Coachman Dr 125-H10 Y
Cobb Cir 124-F10 NN
Cobble Stone 90-C6 Y
Cobbler Way 124-J6 NN
Cobham Hall Cir 124-H6 NN
Cobia Dr 125-J3 Y
Coburn Ct 107-K5 Y
Cockletown Rd 108-K10; 109-A9, B9 Y
Cofer Ct 142-J10 H
Coffman Cir 143-G10 H
Coinjock Run 142-C3 Y
Cokes La 56-E10 J
Colbert Ave 159-E5; HE-D2, E2 H
Colberts La 157-G3 NN
Colberts Trace 109-E8 Y
Colby Rd 72-D10 J
Cold Storage Rd 175-J10 NBN; 91-A6 Y
Cole Dr 89-C9 W
Cole La 140-K9 NN
Cole St 123-H7 F
Colebrook Dr 159-H5 H
Coleman Dr 92-K9 G; 90-D8 J
Colgate Cir 144-C10 H
Coliseum Dr 158-H2 H
Colleen Dr 124-A8, B9 NN; 126-D10 Y
College Blvd 140-J8 NN
College Cr Pl 105-F1 J
College Pl 159-G7; HE-F6 H
College Rd 73-E10; 89-D1 Y
College Terr 89-F6; WE-A2, B3 W
Collette Ct 140-E5 NN
Collier Dr 142-J7 H
Collins Dr 141-G9 NN
Collins La 142-G3 Y
Collinwood Cir 157-K3 H
Collinwood Pl 140-G2 NN
Coloma Dr 124-B8 NN
Colombia Dr 123-H9 NN
Colonel Philip 106-A3 J
Colonel Way 106-A3 J
Colonial Acres Dr 160-C2 H
Colonial Ave 159-A8 H; 90-B6, B7 Y
Colonial Natl Hist Pkwy 104-F5; 105-B6, J1 J; 89-H8; WE-D4, E2 W; 90-F5; 91-A6; 92-B8; 108-E1; YE-A1 Y
Colonial Pl 157-E3 NN
Colonial St 89-J7; WE-D3 W
Colonies Landing 159-G2 H
Colonnade Ct 157-H3 H
Colony Cir 90-C9 J
Colony Dr 89-E5 Y
Colony Pines Dr 124-J6 NN
Colony Pt Rd 105-C1 J
Colony Rd 140-B3, F2, G2 NN
Colony Sq Ct 140-G2 NN
Colony Tr 54-E5, E8 NK
Coltraine Cir 159-J2 H
Columbia Ave 159-D7; HE-B8, C8 H
Colyer Rd 92-H2 G
Comfort Cove 140-K5 NN
Commander Dr 159-A3 H
Commanders Ct 107-C5 J
Commerce Cir 141-G4 Y
Commerce Dr 158-H4 H; 158-C10 NN
Commercial Pl 141-A4 NN
Commodore Dr 143-H9 H; 157-D2 NN
Commons Way 89-F5, F6; WE-A1, B1 Y
Commonwealth Dr 141-F3 Y
Compass Cir 143-J9 H
Compass Ct 141-E1 Y
Compton Ct 142-B8 H
Compton Dr 89-D6; WE-A3 W
Compton Pl 140-J7 NN
Comte De Grasse St 108-K3; YE-D3 Y
Concetta St 158-J10 H
Concord 88-H4 J
Concord Cres 141-D8 NN
Concord Dr 158-E3 H
Condon Rd 123-C8 F
Condor St 159-D6; HE-B4 H
Cone Dr 158-J5, K5 H
Congress Ave 159-E8 H
Congressional Way 88-F5 J
Conies Run 105-K2 J
Conner Dr 159-J3 H
Connie St 160-B4 H
Connors Dr 124-C10 NN
Conservancy Rd 105-F3 J
Constance Ave 104-H5 J
Constance Dr 141-H7 NN
Constant Rd 144-F8 H
Constitution Dr 125-G5 Y
Constitution Dr N 125-G4 Y
Constitution Way 124-D1 NN
Continental Dr 158-J6 H
Convention Dr 158-H4 H
Converse Ct 124-B10 NN
Conway Ct 141-J4 Y
Conway Dr 89-C10; 105-C1 J
Conway Oysterhouse Rd 92-F5 G
Conway Rd 140-J10 NN
Cook Ave 143-D8 L
Cook Rd 108-J4, K6; 109-A7; YE-D6 Y
Cooks Cir 160-B1 H
Cool Ct 108-G4; YE-A4 Y
Cooley Rd 88-K9; 89-A9 J
Cooper Ct 140-F3 NN
Cooper St 159-F5; HE-E2 H
Coopers La 125-F1 Y
Copeland Dr 158-G8 H
Copeland La 157-D2 NN; 109-C10 Y
Copley Pl 140-B1 NN
Copolymer Rd 107-B10 J
Copper Kersey Dr 157-J2 H
Copper Run Rd 106-K6 J
Copperfield 90-C7 Y
Copperfield Rd 158-E3 H
Copse Way 90-E4 Y
Coral Ct 141-A5 NN
Coral Key Pl 141-B5 NN
Coral Pl 159-H4, J4 H
Corbette Pl 125-G3 Y
Corbin Dr 142-H10 H; 140-K9 NN
Cordova Dr 158-A4 H
Corey Cir 159-K6 H
Corinthia Dr 123-K9 NN
Corn Dog Pl 90-E3 Y
Cornelius Dr 158-G5 H
Cornell Dr 159-E5 H; 124-A9 NN
Cornell U Rd 141-C5 NN
Cornwall 88-C2 J
Cornwall Terr 157-K4 H
Cornwallis Pl 124-H7 NN
Cornwallis Rd 109-B5 Y
Corporate Dr 124-K10 NN
Corsair Dr 142-C8 Y
Corwin Cir 142-F7 H
Costigan Dr 124-A10 NN
Cottage Cove La 90-J8 Y
Cottage La 107-H9 NN
Cotton Wood Ave 158-H10 H
Cottonwood Ave 174-H1 H
Cottonwood St 124-F5 NN
Counselor La 144-A9 H
Counselor Way 89-G8 W
Countess Ct 144-A10 H
Country Club Ct 88-J3 J
Country Club Dr 88-K3 J
Country Club Rd 156-K1; 157-A2 NN
Country Dr 39-K4 NK
Country La 125-B2 Y
County Dr 125-F1 Y
County St E 159-J8 H
County St W 159-H7 H
Court De-Ayloon 143-B1 Pq
Court House Dr 124-D10 NN
Court House Green 124-E8 NN
Court Rd 109-C9 Y
Court St 89-H8; WE-C4 W
Courtland Ave 158-J9 H
Courtney Ave 141-G8 NN
Courtney Dr 159-D3 H
Cove Cres 126-A7 Y
Cove Ct 54-G9 NK; 109-E7 Y
Cove Ct S 73-F1 J
Cove Dr 110-C10 Y
Cove Pt La 89-F10 W
Cove Rd 105-G2 J; 123-J9 NN; 143-J3 Pq
Cove Rd N 73-F1 J
Covenant Ct 158-A6 H
Coventry Blvd 141-H4 Y
Coventry La 140-E4 NN
Coventry Rd 88-K10; 89-A10 J
Covina St 107-J10; 123-H1 NN
Cowgirl La 124-E9 NN
Cowles Cir 126-G9 Pq
Cowpen Neck Rd 58-K8 G
Cox Landing 124-B6 NN
Crafford Rd 107-K6; 108-A7 NN
Craggy Oak Ct 72-F1 J
Craig Ave 157-G7 NN
Cranberry St 159-D5; HE-C3 H
Cranbrook Ct 140-H5 NN
Crandol Ct 142-K1 Pq
Crandol Dr 125-G9 Y
Crane Cir 124-A5 NN
Cranefield Pl 140-C4 NN
Cranes Cove 159-A6 H
Cranstons Mill Pond Rd 71-K4 J
Cranwood Ct 141-K2 Y
Crape Myrtle Rd 92-G2 G
Crate Ct 124-K1 Y
Craven St 158-H10 H
Crawford Rd 108-C7, F5; YE-A5 Y
Cream Puff La 108-A4 Y
Creasy Ave 141-H9 NN
Creedmoor Ct 90-E10 Y
Creek Ave 159-F5; HE-D1 H
Creek Blvd 141-E1 Y
Creek Cir 126-G2 Y
Creek Landing 144-B8 H
Creek Pt Cir 73-H2 Y
Creek Rd 92-E4 G
Creek Terr 126-D9 Y
Creek View La 143-K9 H
Creekmere Cove 107-G9 NN
Creekpoint Cove 107-G9 NN
Creeks End Rd 54-G3 NK
Creekshire Cres 107-G9 NN
Creekside Pl 159-H4 H
Creekstone Dr 107-G9 NN
Cremona Dr 124-A10 NN
Crenshaw Ct 158-K5 H
Crepe Myrtle Dr 125-A1 Y
Crescent Ct 142-A5 Y
Crescent Dr 72-D2 J
Crescent Way 124-F9 NN
Cresent Dr 174-K1 H
Crest Lake Ct 141-K2 Y

Crestmont Pl 140-G10 NN
Crestview Dr 72-B1 J
Crestwood Cir 159-E4 H
Crestwood Ct 109-B10; 125-B1 Y
Crestwood Dr 105-G1 J; 141-K6 NN
Crew House Rising 159-G2 H
Crewe Rd 92-H3 G
Crimson Ct 141-J4 Y
Criner La 142-F1 Y
Crispell Ct 158-G1 H
Cristal Dr 124-F4 NN
Criston Dr 141-B3 NN
Crittenden La 140-H10, J10 NN
Critzos Ct 143-F10 H
Croaker Cir 57-A7 J
Croaker Landing Rd 57-C3 J
Croaker Rd 56-G10, K8; 57-A5; 72-F1 J
Croatan Rd 141-B8 NN
Crocker Pl 106-B2 J
Crockett Ct 143-F10; 159-F1 H
Crockett Rd 110-D10; 126-D1 Y
Crocus Rd 92-G3 G
Cromer Ct 159-A8 H
Cromwell Ct 140-H10 NN
Cromwell La 90-D7 Y
Cronin Dr 159-K3 H
Crooked Stick 88-D10 J
Crosby Cir 142-H2 Pq
Crosland Ct 124-E7 NN
Cross Branch 56-A1 J
Cross Cr Rd 72-F1 J
Cross Rd 156-J2 NN
Crossbow Ct 140-E5 NN
Crosscut Ct 56-J8 J
Crossings Ct 124-F9 NN
Crossover Rd 54-H3; 104-B2 J
Crosspointe Ct 141-E1 Y
Croswell Rd 157-K8 NN
Crown Ct 159-H4 H; 124-E5 NN; 90-E4 Y
Crown Pt Dr 141-B2 NN
Crownpoint Rd 105-H2 J
Cruiser Ave 175-H10 NBN
Crump La 89-G8; WE-C4 W
Crutchfield Dr 124-F10; 140-F1 NN
Crystal La 58-D1 G
Crystal Lake Dr 125-J6 Y
Cub Ct 141-H2 Y
Cub La 124-K9 NN
Culotta Dr 142-K10 H
Culpepper Ave 140-K8; 141-A8 NN
Cumberland Ave 159-E9 H
Cumberland Dr 124-F8 NN
Cummings Ave 159-J7 H
Cunningham Dr 158-G3 H
Cure Cir 157-K1 H
Curle Rd 159-F4 H
Curles Cir 104-A2 J
Curls Neck Ct 89-E9 W
Curry Dr 73-B7 J
Curry St 159-J8 H
Curson Ct 126-K10 Pq
Curtail La 70-D3 CC
Curtin Ct 158-G1 H
Curtin Rd 91-A3 Y
Curtis Dr 107-J10; 123-J1 NN
Curtis La 143-E10 H
Curtis Tignor Rd 124-A8, B8 NN
Cushing Post Ct 143-G10 H
Custer Ct 158-C7 H
Custer Pl 124-A8 NN
Custis Dr 90-D9 Y
Cybernetics Way 141-F2 Y
Cynthia Ct 124-D4 NN
Cynthia Dr 158-C2 H
Cypress Cr 88-J5 J
Cypress Crest Dr 92-G4 G
Cypress Crest La 92-G4 G
Cypress Crossing 125-A1 Y
Cypress Ct N 124-B5 NN
Cypress Ct S 124-B5 NN
Cypress Dr 54-G10; 70-H1 J; 54-A7 NK
Cypress Pt Rd 86-J9 CC
Cypress St 159-H5; HE-H2 H
Cypress Terr 124-C6 NN

D

D Ave 124-J8 NN; 109-H7 Y
D St 91-C4 Y
Dabney Dr 140-F4 NN
Dafia Dr 158-D2 H
Dahlia Ct 124-B6 NN
Dahlia La 160-A6 H
Daingerfield Rd 106-B1 J
Daisy Cir 124-C10 NN
Daisy La 141-K3 Y
Dale Ave 175-K8 NBN
Dale St 158-J9 H
Dallas Ct 159-K1 H
Dallas Dr 124-F9 NN
Dam Lake Ct 89-B2 W
Dana Cir 140-B3 NN
Dana Rae Ct 158-B7 NN
Dandy Haven La 110-C7 Y
Dandy Haven Rd 144-C6 H
Dandy Loop Rd 110-B5, B6, B7 Y
Dandy Pt Rd 144-D5 H
Dandy View La 110-B5 Y
Dane Ct 158-A2 H
Danfield Cir 157-K2 H
Danforth Ave 143-D8 L
Daniel Dr 141-G9 NN
Daniel La 160-B4 H
Daphia Cir 141-J5 NN
Daphne Dr 125-H7 Y
Darby Ave 160-A5 H
Darby Ct 160-A9 FM
Darby Rd 125-E9, F10 Y
Darcy Pl 123-G5 F
Darden Dr 124-B10; 140-B1 NN; 126-E7 Pq
Dare Ave 159-A8 H
Dare Rd 125-F5, K3 Y
Darlene La 124-B7 NN
Darnaby Ave 158-J10 H
Darrington Ct 141-J7 NN
Dartmoor Ct 88-K10; 104-K1 J
Dartmoor Dr 124-F8 NN
Darville Dr 159-J3 H
Dastardly La 124-D9 NN
Davenport Dr 158-F2 H
David Cir 140-F5 NN
Davids Way 125-G3 Y
Davis Ave 141-E10; 157-E1 NN
Davis Cir 125-H1 Y
Davis Dr 90-C9 J
Davis Pk Dr 157-K8 NN
Davis Rd 91-C3 Y
Davis Rd E 158-F6 H
Dawn La 141-K7 H
Dawn Pl 142-A2 Y
Dawn Terr 157-H1 NN
Dawson Cir 89-F7; WE-B3 W
Dawson Cres 110-A10 Y
Dawson Dr 110-A10 Y
Day Cir 123-K9 NN
Day One Ct 124-H9 NN
Day St 158-H10 H
Daybreak Cir 124-H8 NN
Daybreak Ct 124-H8 NN
Daylight Ct 124-H8 NN
Days Neck Rd 155-B6 I
Days Pt La 155-C6 I
Days Pt Rd 155-B8, D7 I
De Alba La 110-E10; 126-D1 Y
De Laura Dr 124-B9 NN
De Wald Cir 140-H2 NN
Deal 88-H3 J
Deal Dr 124-F6 NN
Dean Ray Ct 157-K8 NN
Deans Cir 140-G10 NN
Deans Rd 40-B7 NK
Deaton Dr 143-G10 H
Deauville Cir 140-H7 NN
Debbie Dr 141-J7 H
Debbie La 140-C1 NN
Deborah La 157-K9 NN
Debra Dr 89-D5 J
Decatur Ave 175-H9 NBN
Decatur Dr 92-F4 G
Decatur St 158-A10 NN
Decesare Dr 157-K2 H
Declaration Dr 125-G5 Y
Declaration Way 124-H8 NN
Deep Cr Rd 140-G8; 141-A8 NN
Deep Run Rd 158-D3 H
Deep Spring Dr 140-C1 NN
Deep Water Cove 141-E8 NN
Deer Path Rd 73-J4 Y
Deer Path Tr 124-G4 NN
Deer Pk Dr 141-D9 NN
Deer Run Tr 140-A1 NN
Deer Spring Rd 88-K9 J
Deer View La 144-B8 H
Deere Cir 57-C7 J
Deerfield Blvd 142-J10 H
Deerfield Ct 88-C10; 104-C1 J
Deerlope Tr 88-G9 J
Deerwood Dr 87-G2 J
DeGaule St 158-C9 NN
DeGrasse St 72-J10 J
Dehardit Ave 92-K10 G
DeHaven Ct 72-J10 J
Dehaven Ct 88-J1 J
Deidre La 140-D1 NN
Del Lago Dr 107-B6 J
DeLafayette Pl 72-E9 J
Delaware Ave 158-J9 H; 89-F6; WE-A2 W
Dell Dr 158-A3 H
Dellwood Dr 140-D4 NN
Delmar La 124-F10 NN
Delmont Cir 142-A7 H
Deloice Cres 124-E10 NN
Delta Cir 157-J2 NN
Deluca Way 141-F1 Y
Demetro Dr 160-A3 H
Dena Dr 88-A6 J
Denbigh Blvd 124-D10, F9, H7; 140-A1 NN; 125-B4 Y
Denbigh Pk Dr 124-C7 NN
Dendron Pl 140-C5 NN
Denise Dr 126-A10 Y
Dennis Dr 90-D4 Y
Denton Dr 160-B5 H
Denver Cir 142-K10 H
Depot St 56-B9 J
DePriest Downs 123-K4 NN
Deputy La 124-D8 NN
Derby Dr 142-G7 H
Derby La 104-C3 J
Derby Run 142-E4 Y
Derosa Dr 157-J1 H
Derry Ct 140-D3 NN
Derry Rd 159-J6 H
Deshazor Dr 124-D10 NN
Deveron Dr 142-J1 Pq
Devils Den Rd 143-F9 H
Devol Dr 123-K9 NN
Devon Pl 140-J7 NN
Devon Rd 88-H2, K2 J
Devonshire Dr 73-E2 Y
Devonshire Terr 158-C3 H
Devore Ave 157-J1 H
Dewey Ave 174-G1 H
Dexter La 143-K7 H
Diamond Dr 41-J4 KQ
Diamond Hill Dr 157-J2, J3 H
Diane Tr 141-H10 NN
Diascund Pointe 54-G6 NK
Diascund Pointe Tr 54-F6 NK
Diascund Rd 54-J4; 55-A6, C9 J
Diascund Res Rd 54-F1 J; 38-F10; 54-F1 NK
Dibby Rd 92-H5 G
Dickman St 123-G5 F
Dickson Cir 55-J5 J
Diesel Dr 90-J7, K6; 91-A7, A8 Y
Diesel Rd 90-K7; 91-A7 Y
Digges Ct 73-B10 J
Digges Dr 140-C5 NN
Digges Rd 91-J8 Y
Diggs Dr 143-A10; 158-K1; 159-A1 H
Diligence Dr 141-F7 NN
Dillard La 125-F10 Y
Dillard St 89-F6; WE-A2 W
Dillingham Blvd 175-H8 NBN
Dillwyn Dr 140-C4 NN
Dilts Dr 124-F6 NN
Dimmock Ave 141-D10 NN
Dinwiddie Pl 124-H7 NN
Diplomat Ct 124-E6 NN
Discovery La 104-H5 J
Discovery Rd 144-F8 H
Dixie Pl 141-F10 NN
Dixon Rd 126-H7 Pq
Dockside Dr 159-F4 H
Doctors Cr Rd 54-B2 NK
Dodd Blvd 143-E8 L
Dodge Dr 90-C8 Y
Doe La 141-F4 NN
Dog Town La 92-H2 G
Doggett La 71-H5 J
Dogleg Dr 88-E3 J
Dogwood Ave 143-B9 L
Dogwood Ct 109-C8 Y
Dogwood Ct N 124-C5 NN
Dogwood Ct S 124-C5 NN
Dogwood Dr 105-C1 J; 157-A3 NN; 142-D4 Y
Dogwood La 124-G3 NN
Dogwood Rd 108-B4 Y
Dogwood St 159-C6 H
Dolphin Ct 144-B10 H
Dolphin Dr 125-K3 Y
Dominion Dr 140-J4 NN
Domino Dr 107-C5 J
Don Juan Cir 141-G2 Y
Don Rett Cir 123-K9 NN
Don-eve Ct 140-E1 NN
Donald St 159-E2 H
Donna Pl 141-A5 NN
Donnelson Pl 123-J6 F
Dooley St 159-C6; HE-A4 H
Doolittle Rd 142-J5; 159-C3 H
Doolittle St 143-E7 L
Dora Dr 140-F2 NN
Doral 88-F4 J
Doral Dr 141-D2 NN
Dorchester Ct 141-J7 NN
Dorene Ct 159-J2 H
Dorene Pl 124-D3 NN
Dorlie Cir 126-J10 Pq
Dornach 88-H5 J
Dorothy Dr 126-F8 Pq; 125-F3 Y
Dorris Carlson Dr 142-C9 H
Dorset Mews 89-B2 J
Dorsetshire Terr 158-C3 H
Dorsey Rd 141-C8 NN
Dotorg Rd 89-C5 J
Dotson Dr 159-E4 H
Douglas Dr 157-D4 NN; 90-E10 Y
Douglas La 105-D2 J
Douglas St 159-K6 H; 143-E7 L
Dove Ct 141-B8 NN
Dover Rd 158-E2 H; 105-B1 J
Downer La 157-J3 H
Downes St 159-J8 H
Downey Farm Rd 158-H1 H
Downey St 91-C5 Y
Downing Ct 89-C3 J
Downing Pl 156-J1 NN
Downing St 159-A7 H
Dozier Rd 123-H4 NN
Dragon Dr 142-C8 Y
Drake Ct 141-K7 H; 140-H8 NN
Drammon Ct 72-F4 J
Draper Cir 157-J1 H
Draper La 156-J1 NN
Dresden Dr 157-G2 NN
Dressage Ct E 142-D6 H
Dressage Ct W 142-D7 H
Drew Rd 90-E10; 106-F1 Y
Drewry La 54-H7 J; 125-J1 Y
Driftwood Ct 141-E1 Y
Driftwood Dr 158-C6 H
Driftwood Way 88-E5 J
Drivers La 141-C2 NN
Druid Ct 89-B10 J
Druid Dr 89-B10; 105-B1 J
Drum Pointe 107-G8 NN
Drummond Ct 88-F2 J
Drummonds Way 143-H9 H
Dryden Dr 126-E8 Pq
Dryden La 126-C9 Y
Dryden St N 142-J5 H
Dryden St S 142-H5 H
Dublin Ct 141-J7 NN
Dubois Dr 140-D1 NN
Duchess Ct 160-A1 H
Duck Commons 72-J9 J
Dudley Dr 105-E2 J
Dudley Rd 91-J9 Y
Duer Rd 89-A9 J
Duff Dr 125-K4 Y
Duffie Dr 90-C8 J
Duke of Gloucester St 89-G7; WE-C3 W
Duke Rd 92-K9 G
Duke St 158-H7 H; 174-D1 NN
Dulcy Ct 158-D4 H
Duller La 123-J6 F
Duluth Ct 158-D7 H
Dumont Cir 160-A1 H
Dunbar Rd 158-H9 H
Duncan 88-J4 J
Duncan Dr 160-A5 H; 90-E8 Y
Duncombe La 124-G4 NN
Dundee 88-K3 J
Dundee Rd 158-J7 H
Dundee Way 123-J4 NN
Dunes 88-H5 J
Dunham Massie Dr 159-H2 H
Dunhill Way 141-D2 NN
Dunkin Rd 54-B2 NK
Dunmore Dr 140-D3 NN
Dunn Cir 157-J2, K2 H
Dunnavant La 140-G9 NN
Dunning St 89-G6; WE-C1 W
Dunnshire Terr 158-C3 H
Dunrovin La 58-J9 G
Dunwoody Cir 157-K3 H
Durand Loop 143-A7 L
Durand St E 142-K6 H
Durand St W 142-J6 H
Durant Tr 107-G8 NN
Durfeys Mill Rd 105-A2 J
Durham Ct 88-J10 J; 142-A5 Y
Durham St 143-K10 H
Dusk Ct 141-K7 H; 141-E8 NN
Dusty La 73-C2 Y
Dutchess Ct 144-A10 H
Dutchess La 124-A8 NN
Duval Ct 143-J10 H
Dwight Rd 141-F10 NN
Dwyer Cir 123-J7 F
Dyke 88-J3 J

E

E Ave 109-J7 Y
E St 158-E9 H; 91-D5 Y
Eads Ct 124-C9 NN
Eagan Ave 143-E8 L
Eagle 88-D3 J
Eagle Ave 142-B5 Y
Eagle La 157-J8 NN; 125-K2 Y
Eagle Nest La 155-C9 I
Eagle Pt Rd 159-C2 H
Eagles Landing 160-A1 H
Eagles Nest Rd 70-D6, G8, G10; 86-G1 CC
Earl Dr 107-C7 J
Earl St 159-D8 H; 125-H1 Y
Earle Ct 124-A10 NN
Earlien Rd 71-C4 J
Early St 90-A8 W
Earnestine Ave 89-D4 W
East Ames St 142-J5 H
East Autumn 72-J10 J
East Ave 159-C9 H
East Bay 104-B1 J
East Bayberry Ct 159-E2 H
East Beaver Run 72-G10 J
East Beechwood Rd 107-H5 Y
East Branch Rd 109-D7 Y
East Bristol La 142-E4 Y
East Brittington 104-C1 J
East Bush Rd 142-J4 H
East Co St 159-J8 H
East Davis Rd 158-F6 H
East Dressage Ct 142-D6 H
East Durand St 142-K6 H
East Fall 72-J10 J
East Gilbert St 159-E3 H
East Governor Dr 140-D6 NN
East Is Rd 103-G1 J
East John Proctor 104-H2 J
East King St 143-E10 H
East Lake Shore Dr 158-F4 H
East Lamington Rd 159-E1 H
East Landing 106-B4 J
East Lewis Rd 158-F6 H
East Manor Dr 90-E8 Y
East Mercury Blvd 159-J6 H
East Miller Rd 39-H9 J
East Pembroke Ave 159-F5, J5; 160-B4; HE-E3, G2 H
East Pinto Ct 142-D6 H
East Preston St 159-E2 H
East Queens Dr 90-E4 Y
East Queens Way 159-E6; HE-E5 H
East Rangeley 104-C2 J
East Reid St 142-K6 H
East Rexford Dr 124-A7 NN
East Roan Ct 142-D6 H
East Rochambeau Dr 73-A6, B5, B6, E6, H9; 89-K1 Y
East Russell Ct 158-C10 NN
East Russell Rd 158-F6 H
East Sandy Pt Rd 126-J7 Pq
East Savannah Sq 89-F4 Y
East Settlers Landing Rd HE-F5 H
East Southampton Ave 159-D8 H
East Spring 72-J10 J
East Steeplechase Way 88-H8 J
East Summer 72-J10 J
East Sunset Rd 159-D8 H
East Taylor St 142-K7 H; 143-B9 L
East Tiverton 104-B1 J
East Troy Dr 141-A5 NN
East Valasia Rd 126-H10 Pq
East Victoria Blvd HE-C7 H
East View Rd 92-H3 G
East Vill Pk Dr 88-J10; 104-H1 J
East Walcott St 142-K6 H
East Walker Rd 158-E5 H
East Weaver Rd 158-F5 H
East Wedgewood Dr 125-D10 Y
East Whittaker Close 103-H1 J
East Winona Dr 158-K9 H
East Winter 72-J9 J
East Woodland Dr 125-K8 Y
East Yorktown Rd 126-E10 Pq
East-West Expressway 141-G10 NN
Eastbriar Ct 158-E7 H
Eastbury 88-H4 J
Easterly Ave 159-D4 H
Eastfield Ct 158-D6 H
Easthill Ct 160-C2 H
Eastlawn Dr 144-C10 H
Eastmoreland Dr 159-G3 H
Eastnor Ct 124-H6 NN
Eastwind Cove 140-K5 NN
Eastwood Ave 159-A7 H
Eastwood Dr 140-E4 NN
Eastwood Homestead La 58-J1 G
Easy St 158-C5 H
Eaton St 159-F5; HE-D2, E3 H
Ebb Cir 38-C3 NK
Ebb Cove Ct 141-C1 NN
Ebb Tide Landing 126-J7 Pq
Ebbing Quay 158-B1 H
Eberly Terr 159-F5 H
Echo Hall Terr 92-J5 G
Echo Ridge Ct 107-G8 NN
Echols Ct 158-D6 H
Eclipse Ct 141-E8 NN
Ed Wright La 141-B5 NN
Edale Ave 90-C8 Y
Eden River 88-F3 J
Edenbrook Dr 157-J2 H
Edgehill La 108-K9 Y
Edgemont Dr 142-A7 H
Edgemoor Dr 107-H9 NN
Edgewater Dr 141-C1 NN
Edgewater Rd 144-B8 H
Edgewood Ct 125-G6 Y
Edgewood Dr 158-E4 H; 140-H8 NN
Edgewood La 105-B1 J
Edinburgh Dr 88-F3, G4, H5 J
Edinburgh La 143-K8 H
Edith Ct 159-F1 H
Edith Key St 158-E2 H
Edith La 38-J9 NK; 126-D4 Y
Edmond Dr 141-B8 NN
Edmonds Cove Rd 144-D8 H
Edna Dr 72-E9 J
Edney Dr 140-G4 NN
Edson Terr 159-K5 H
Edsyl St 140-F3 NN
Edward Wakefield 106-B2 J
Edward Wyatt Dr 88-F1 J
Edwards Ct 124-C4 NN; 142-F4 Y
Edwards Dr 158-B5 H
Edwards Rd 126-E6 Pq
Edythe St 174-A1 NN
Effingham Pl 124-H7 NN
Efford Ct 88-G1 J
Egger Cir 160-A3 H
Eggleston Ave 159-D9 H
Eisele Ct 157-J1 H
Eisenhower Dr 110-C7 Y
El Dorado Ct 143-G10 H
El Passo Ct 143-J10 H
El Sombrero Ct 107-B6 J
Elaine Dr 140-E1 NN
Elbow Beach La 107-B6 J
Elder Rd 123-J9 NN
Eleanor Ct 124-G9 NN
Electra Dr 124-K9 NN
Elezabeth Lake Dr 159-H2 H
Elise Pl 141-H4 Y
Elizabeth Champion Ct 87-J10 J
Elizabeth Ct 158-B8 NN
Elizabeth Killebrew 88-B6 J
Elizabeth Lake Dr 159-H3 H
Elizabeth Meriwether 106-A2 J
Elizabeth Page 105-K3 J
Elizabeth Rd 159-D9 H
Elizabeths Quay 126-B5 Y
Elksnin La 125-K1 Y
Ella Taylor Rd 125-H8 Y
Ellen Rd 158-A10 NN
Ellerson Ct 125-H1 Y
Elliffe Rd 157-H1 NN
Ellington Ave 158-F10 H
Ellinson Ct 106-B3 J
Elliott Pl 144-D8 H
Elliott Rd 142-E1 Y
Ellis La 92-G2 G
Ellyson St 175-K10 NBN
Elm Ave 159-F5; HE-E2 H; 157-D4 NN
Elm Ct N 124-B5 NN
Elm Ct S 124-B5 NN
Elm Dr 125-B7 Y
Elm Lake Way 142-A3 Y
Elm St 92-H2 G; 143-A9 H
Elmhurst Dr 125-B1 Y
Elmhurst St 107-H10, J9 NN
Elmwood La 158-A1 H; 56-H7 J
Elowro Dr 140-C4 NN
Elsie Dr 124-F2 NN
Eltham 88-C3 J
Eltham Ct 159-J2 H
Eltham Rd 38-F2 NK
Elton Hall Cir 124-H5 NN
Emancipation Dr 159-G7; HE-H6 H
Embers La 107-B6 J
Emerald Ct 124-F4 NN; 142-A3 Y
Emeraude Pl 158-C5 H
Emerson Cir 140-F2 NN
Emma Dr 144-C10 H; 158-C8 NN
Emma Kate Ct 158-C7 NN
Emmaus Rd 126-E7, F8 Pq
Emmaus Rd S 126-G8 Pq
Emmett La 158-D4 H
Emmonds Rd 143-D6 L
Emory Pl 126-E9 Pq
Empire Ct 124-F9 NN
Empress Ct 107-B5 J
Emrick Ave 141-D10 NN
Enchanted Forest La 141-K3 Y
Endeavor Dr 107-B7 J
Ends Rd 58-J1 G
Enfield Dr 158-H3 H
Engineers Rd 160-A10 FM
England Ave 159-C7 H
England Cir S 105-J1 J
England Rd 92-K9 G
England St N 89-H7; WE-D2 W
England St S 105-J1 J; 89-J7; WE-D4 W
Englewood Dr 125-A1 Y
Enos Ct 124-F3 NN
Enscore Ct 157-H2 H
Ensign Dr 143-H9 H
Ensigne Spence 106-C2, D2 J
Enterprise Dr 123-F2, J2 NN
Enterprise Pkwy 158-E2, F1 H
Eppington Cir 159-J2 H
Ericcson Dr 159-E9 H
Erin La 157-K9 NN
Erin Leigh Cir 140-G5 NN
Erin Leigh Ct 90-C6 Y
Ernest La 89-A7 J
Erskine St 158-C2 H
Ervin St 159-B8 H
Escort Ave 175-H10 NBN
Essex Pk Dr 144-B10; 160-B1 H
Estate Ct 55-F5 J
Estate Dr 157-J3 H
Estates Ct 54-C4 NK
Estates Dr 54-C4 NK
Estates La 54-C4 NK
Esteban Ct 87-H2 J
Estelle Ct 124-E2 NN
Esterdale La 143-E10 H
Ethel Dr 158-D8 H
Etienne Ct 87-K5 J
Etna Dr 123-K8 NN
Eton Cove 124-G6 NN
Eubank Cir 141-H7 NN
Eubank Dr 158-D2 H
Eureka Loop 157-G4 NN
Eustis La 159-K9 FM
Eva Ct 141-G6 NN
Evan Russell St 92-H2 G
Evans Cir 127-A8 Pq
Evans Grove Dr 127-D10 Pq
Evans St 158-G6 H
Evelyn Ct 157-K6 H
Evelyn Dr 140-A1 NN
Everett Dr 140-C5 NN
Evergreen Cir 110-D10 Y
Evergreen Pl 158-J1 H
Evergreen Way 104-A1 J
Eves St 91-J8 Y
Evesham Ct 104-K1 J
Ewell Cir 89-G7; WE-C4 W
Ewell Ct 158-G7 H
Ewell Pl 88-G7 J
Ewell Rd 73-C9 Y
Excalibur Pl 141-F5 NN
Executive Dr 158-G1 H; 140-J9 NN
Exeter Ct 158-B2 H
Exeter Rd 140-D4 NN
Exmoor Ct 89-D10 W
Export Cir 157-E4, F3 NN
Exter Ct 105-A1 J

F

F Ave 124-J8 NN; 109-H7 Y
F St 91-E5 Y
Fair Chase 104-F3 J
Fairchild Ct 157-J3 H
Fairfax Ave 158-B10; 174-B1 NN
Fairfax Dr 159-B9 H
Fairfax Way 105-J4, K4; 106-A4 J
Fairfield Blvd 144-A10; 160-A1 H
Fairfield Ct 140-D3 NN
Fairfield Dr 126-A6 Y
Fairland Ave 159-A6, B7 H
Fairlane La 160-A5 H
Fairmont Dr 142-A7 H; 56-E9 J
Fairview Dr 56-F9 J
Fairview Rd 156-K1 NN
Fairway La 140-K10 NN; 141-E2 Y
Fairway Lookout 88-D10 J
Fairways Reach 88-D10 J
Fairwinds Dr 141-E1 Y
Faith Ave 140-G3 NN
Falcon Dr 140-G6 NN; 142-C8 Y
Falcon Rd 109-A8 Y
Falkirk Mews 89-B2 J
Fall E 72-J10 J
Falling Cr Cir 104-A2 J
Falling Spring Run 125-D4 Y
Fallmeadow Ct 157-K3 H
Fallon Ct N 159-B7 H
Fallon Ct S 159-B7 H
Falls Reach Pkwy 107-G9 NN
Falmouth Cir 141-C2 NN
Falmouth Turning 143-H10 H
False Bay Pl 109-D7 Y
Far St 142-K1; 143-A1 Pq
Farm House La 143-J8 H
Farm Rd 155-E5 I
Farm St 159-C4 H
Farmers Dr 39-G6, J1 NK
Farmington Blvd 157-K3 H
Farmstead Pl 125-A1 Y
Farmville La 72-F2, G2 J
Farrington Pl 160-A4 H
Fast Cir 159-K2 H
Faubus Dr 158-B8 NN
Faulk Cir 159-J3 H
Faulkner Rd 125-C2 Y
Fauquier Pl 124-H7 NN
Fawn Cir 158-C2 H
Fawn La 141-F4 NN
Fawn Lake Dr 124-G4 NN
Fawn Pl 87-G1 J
Fay Cir 124-A8 NN; 141-J1 Y
Feathergrass Pk 125-E4 Y
Felgates Rd 91-F8 Y
Felton Pl 158-K3 H
Fenton Mill Rd 56-J8, J9; 72-K1; 73-A1 J; 73-A1, D5 Y
Fenwick Rd 160-A10, B9, C6 FM
Fenwood Cres 124-B10 NN

Fenwyck Ct 88-H2 J
Ferguson Ave 157-E5, F6 NN
Ferguson Bend 141-K4 Y
Ferguson Cove 140-G10; 156-G1 NN
Ferguson La 141-E10; 157-E1 NN
Ferguson St 126-E7 Pq
Fergusson Cir 123-G7 F
Fern Cove Ct N 142-A2 Y
Fern Cove Ct S 142-A2 Y
Fern Ct 88-G1; 105-H1 J
Fern Dr 141-C10 NN
Fern St 158-H10 H
Ferncliff Dr 89-B8 J
Ferndale Dr 140-K8 NN
Fernwood 103-K1 J
Fernwood Bend 125-F5 Y
Ferrell Dr 107-B6 J
Ferrier Pl 142-A4 Y
Ferry Landing La 160-A2 H
Ferry Rd 159-F7; HE-E6 H
Fidd Rd 90-K5; 91-A5 Y
Fiddlers Green 159-E8 H
Fiddlers Ridge Pkwy 104-B1 J
Fieldcrest Ct 88-C10 J
Fielding Lewis Dr 125-J4 Y
Fields Dr 160-C2 H
Fields Landing Rd 92-F2 G
Fieldstone Ct 141-E1 Y
Fieldstone La 124-G10 NN
Fieldstone Pkwy 55-H2 J
Fig Tree Rd 108-A3 Y
Fillmore Dr 90-D9 Y
Filly Run Rd 58-K1 G
Filson Ct 124-G5 NN
Fincastle Dr 141-J7 NN
Finch La 109-K9 Y
Finch Pl 124-H6 NN
Findley Sq 158-K5; 159-A5 H
Findley St 159-A5 H
Finland St 158-H9 H
Finns Pt La 144-A7 H
Finns Rd 54-B3 NK
Fir Ct 158-B5 H
Firby Rd 125-G9 Y
Fire Tower Rd 39-H10; 55-G1 J
Firestone 88-F6 J
Firethorn Pl 105-H1 J
Firth La 126-H10 Pq
Fiscella Ct 143-F9 H
Fischer Dr 140-D2 NN
Fisherman Cove 126-A2 Y
Fishers Landing Rd 140-G9 NN
Fishing Pt Dr 141-C6 NN
Fishneck Landing Rd 126-C3 Y
Fithian La 88-G8 J
Fitzhugh Dr 140-J4 NN
Five Forks La 143-G10 H
Five Pillars La 88-H7 J
Flag Cr Rd 126-D9 Y
Flag Stone Way 124-F3 NN
Flagship Dr 124-E3 NN
Flamingo Pl 126-A3 Y
Flat Rock Rd 157-F7 NN
Flax Mill Rd 140-D7 NN
Fleetwood St 159-E8 H
Fleming Cir 141-A10 NN
Fleming Dr 159-F3 H
Fletcher Rd 41-E4 KQ
Flint Dr 124-F10 NN
Flintlock Dr 89-B7 Y
Flinton Dr 142-B6 H
Flora Dr 124-C10 NN
Florence Dr 158-C4 H
Florida St 159-B6 H
Flowers Terr 124-B5, B6 NN
Floyd Ave 143-C1 Pq
Floyd Thompson Blvd 142-G7 H
Floyd Thompson Dr 142-J7 H
Flume Run 141-A2 NN
Fluvanna Rd 141-J7 NN
Flyer Dr 142-B5 Y
Foley Dr 90-C8 J
Foley St 159-D5 H
Folsom Ct 140-H8 NN
Fontaine Rd 157-D2 NN
Ford Ct 124-A10 NN
Ford Rd 160-A4 H
Fordham La 160-A5 H
Fords Colony Dr 72-F10; 88-E2 J
Forest Ct 72-D10 J
Forest Dr 92-K9 G
Forest Glen Dr 72-D9 J
Forest Hgts Rd 73-A8 J
Forest Hill Dr 89-K5; WE-F1 W
Forest Hollow La 88-H9 J
Forest La 73-K4 Y
Forest Lake Rd 54-G8 J
Forest View Rd 124-C8 NN
Forge Rd 55-A10, H10; 71-E1 J
Formby 88-G4 J
Forrest Dr 141-D8, E8 NN; 127-B9 Pq
Forrest Rd 127-A9, A10 Pq
Forrest St 159-E1 H
Forrestal Cir 124-C4 NN
Fort Ave 159-H7 H
Fort Boykin Tr 155-B5 I
Fort Eustis Blvd 123-J3; 124-F2 NN; 109-B10; 124-F2, H2; 125-A1 Y
Fort Fun Cir 157-G7 NN
Fort James Ct 54-F6 NK
Fort Worth St 143-J10; 144-A10; 159-K1 H
Forth River 88-F3 J
Forum St 159-F1 H
Fosque La 92-F3 G
Foster Myer La 92-K6 G
Foster Rd 89-B4, C4 J
Founders Hill N 103-B1 J
Founders Hill S 103-B1 J
Founders Way 103-C1 J
Foundry Ct 142-A3 Y
Fountain Trace 142-C5 Y
Four Islands Tr 54-F9 NK
Four Mile Tree 58-A9 J
Fowler Ave 174-F1 H
Fowlers Close N 87-D10 J
Fowlers Lake Rd 87-D10 J
Fox Den 72-E8 J
Fox Gate Way 144-D10 H
Fox Grove Dr 160-C1 H
Fox Hill Rd 159-F2; 160-A1 H; 72-E8 J
Fox Hollow 72-E8 J
Fox Hunt Tr 72-E8 J
Fox La 110-B9 Y
Fox Run 158-K2 H; 72-D8 J
Fox Wood La 144-C7 H
Foxboro Dr 141-A1 NN
Foxchase Way 89-D6 J
Foxcroft Rd 88-K9 J
Foxfield Pl 125-K4 Y
Foxhaven Dr 58-J9 G
Foxridge Rd 88-H9 J
Foxtown Rd 126-F9 Pq
Foxwood Dr 125-F10 Y
Fran Cir 159-B7 H
Frances Berkeley 88-A7, B7 J
Frances St 141-E10 NN
Frances Thacker 106-B4 J
Francis Chapman N 104-G1 J
Francis Chapman W 104-G1 J
Francis Cir 125-G9 Y
Francis Dr 125-J7 Y
Francis Jessup 106-B2 J
Francis St 89-H7; WE-C4 W
Francisco Way 157-G4 NN
Frank Hunt Ct 142-G2 Pq
Frank La 159-K9 FM; 140-H10 NN
Frankie P Engle Cir 158-D2 H
Franklin Blvd 159-F8 H
Franklin Ct 142-A3 Y
Franklin Rd 108-K1 G; 157-G7 NN; 126-B5 Y
Franklin St 159-E6; HE-C4 H; 89-J7; WE-E3 W
Franktown Rd 160-B2 H
Fraser La 92-J8 G
Fraser Rd 92-J8 G
Frazee Ct 92-G3 G
Fred Johnson Jones 106-A3 J
Freda Ct 158-D7 H
Frederick Dr 92-H5 G; 157-E1 NN
Freedom Blvd 125-D2 Y
Freedom Cir 157-K5 H
Freedom Dr 107-B7 J
Freeman Dr 158-K4; 159-A5 H; 127-C10 Pq
Freeman La 142-H3 Pq
Freeman Rd S 103-G1 J
Freemans Trace 141-J2 Y
Freemoor Dr 126-E8 Pq
Frenchmans Key 89-F10 W
Friant La 92-E1 G
Friar Cir 140-G4 NN
Friars Ct 90-F5 Y
Friedman Pl 124-G6 NN
Friendly Dr 157-J5 H
Friendship Dr 72-A3 J
Friendship Rd 124-J9 NN
Frigate Dr 124-D3 NN
Frissell Ave 159-G9; HE-F7 H
Frissell St 159-H7 H
Frond Ct 88-K8 J
Frost Ct 158-J5 H
Fruitwood Dr 158-C5 H
Frying Pan Farm La 92-F4 G
Fuel Dr N 90-K6; 91-A6 Y
Fuel Dr S 90-K6; 91-A6 Y
Fuel Farm Rd 109-B5 Y
Fuller La 160-C7 FM
Fullinwinder La 108-C3 Y
Fulton Ave 159-J7 H
Fulton Farm Rd 159-G4 H
Fundy Ct 124-G4 NN
Fyfe Ct 89-G6 Y

G

G Ave 124-K10; 140-K1 NN; 109-G7 Y
G Maria Dr 126-J8 Pq
G St 158-G9 H
Gabriel La 124-E2 NN
Gaelic Ct 140-K5 NN
Gaffey Pl 123-E4 F
Gail Ct 158-C8 NN
Gaines Mill La 143-H10 H
Gaines Way 125-J5 Y
Gainsborough Pl 124-H6 NN
Galahad Dr 124-G7 NN
Galax St 158-F8 H
Galaxy Way 142-B1 Y
Galena Tr 70-G4 CC
Gallaer Ct 142-E7 H
Galleon Dr 124-D3 NN
Gallo Ct 89-G6; WE-B1 Y
Gallop Pl 124-E3 NN
Galt Dr 89-B4 J
Galveston Ct 143-K10 H
Gambol St 157-F2 NN
Garden Dr 174-F4 NN
Garden State Dr 140-H2 NN
Gardenville Dr 141-H3 Y
Gardner Ct 89-B1 J
Gardner La 92-E5 G
Garfield Dr 124-A8 NN
Gargamel Rue 123-J2 NN
Garland Ct 141-A9 NN
Garland Dr 140-K9; 141-A9 NN
Garland St 159-G6; HE-G3 H
Garner Terr 174-F3 NN
Garnett Cir 140-G3 NN
Garrett Dr 158-H7 H; 89-B3 J
Garris Dr 158-G2 H
Garrison Dr 89-E5; WE-A1 W
Garrison Rd 91-B6, D6, E5 Y
Garrow Cir 159-K2 H
Garrow Rd 124-A10, B10; 140-A2 NN
Gary La 158-K8 H
Gary Rd 157-D4 NN
Gaston Ct 141-D8 NN
Gate House Blvd 104-K4 J
Gate House Rd 123-J3 NN
Gate St 140-E4 NN
Gates Rd 40-B2 NK
Gateway Blvd 158-E1 H
Gateway Ct 141-C1 NN
Gateway Dr 159-B1 H
Gatewood Ct 159-F7; HE-F6 H
Gatewood Rd 157-D1 NN
Gatewood St HE-F6 H
Gatling Dr 158-F6 H
Gawain Way N 159-E4 H
Gawain Way S 159-E4 H
Gay Dr 141-B9 NN
Gay Lynn Dr 142-H1 Pq
Gayle St 159-J5 H
Gaylor La 124-E10 NN
Geddy Dr 158-J6 H
Geist Rd 144-D8 H
Gemini Way 142-G5 H
Gena Ct 140-D2 NN
General Ct 124-E8 NN
General Gookin Ct 104-C2 J
Genneys Way 140-G1 NN
Genoa Dr 160-D1 H
Gentry La 72-K10 J
George Ct 159-J4 H
George Emerson La 141-G4 Y
George Mason 88-C7 J
George Perry 106-A2 J
George Washington Mem Hwy 92-H2 G; 108-H5; 109-B9; 125-C2, F5; 141-G1; YE-B6 Y
Georgeanna Ct 159-K6 H
Georgetown Cres 89-E4 Y
Georgia Ct 140-K4 NN
Georgia St 159-B6 H
Gersham Pl 104-J5 J
Gibson Dr 159-E3 H
Gilbert Dr 90-C9 J
Gilbert St 159-C2 H; 175-H9, K9 NBN
Gilbert St E 159-E3 H
Gildner Rd 158-J4; 159-A4 H
Giles Bland 88-B7 J
Giles La 108-G5; YE-A6 Y
Gilford Ct 124-D3 NN
Gilley Dr 89-A9 J
Gilliam Rd 107-E2 Y
Ginger Ct 73-A10 J
Ginger Tr 124-E6 NN
Gingerwood Ct 124-B10 NN
Ginny Hill Rd 58-F5 G
Giovanni Ct 140-H2 NN
Glade Rd 141-C9 NN
Gladstone Ct 158-D7 H
Gladys Dr 88-K9 J
Glascock Ct 142-J10 H
Glascow Way 159-K2 H
Glasgow 88-K3 J
Glebe Spring La 125-K9 Y
Glen Allen Ct 107-H9 NN
Glen Cove 126-D1 Y
Glen Forest Dr 143-J10 H
Glen Laurel Way 125-E5, F5 Y
Glen More Dr 142-J1 Pq
Glendale Rd 159-A8 H; 141-B9 NN
Gleneagles 88-F3 J
Gleneagles Way 125-C10 NN
Glenhaven Dr 144-C10; 160-C1 H
Glenkinchiey Ct 141-E1 Y
Glenn Cir 90-C7 Y
Glenn La 141-B10 NN
Glenrock Dr 158-F7 H
Glenwood Dr 105-H1 J
Glenwood Rd 159-F1 H
Glenwood Rd N 159-E1 H
Glica Ct 157-J1 H
Gloria Ct 142-J7 H
Gloucester Dr 158-A10 NN
Gloucester Rd 92-K9 G
Gloucester St 158-H10; 174-G1 H
Glover Ave 143-D8 L
Glynis Bell 104-H2 J
Gnarled Oak La 125-E4 Y
Godfrey Dr 140-B1 NN
Godspeed La 104-H5 J
Goering Dr 73-J2 Y
Gold Cup Rd 158-D3 H
Golden Gate Dr 160-A2 H
Golden Willow Cir 158-K1 H
Goldmont Cir 142-A7 H
Goldsboro Dr 157-J5 H
Goldsmith Pl 141-C8 NN
Golf Club Rd 106-D4 J
Gooch Dr 89-F7; WE-A3, B3 W
Good Hope Rd 38-A4, D6 NK
Goode Dr 140-E2 NN
Goodman Rd 123-F7 F
Goodrich Durfey 105-D3 J
Goodwin Neck Rd 109-F9; 110-B7; 125-E2 Y
Goodwin Rd 140-J10 NN
Goodwin St 89-F8 W
Goose Cir 124-A5 NN
Goose Cr Rd 110-B10 Y
Goosley Rd 108-F4; YE-A5, C5 Y
Gordon Ct 158-A5 H
Gordon Dr 126-G9 Pq
Gordon La 141-H3 Y
Gorham Ct 125-C10 Y
Gosnold Pl 140-J6 NN
Gothic Rd 86-A3 CC
Gourmand Rd 103-A10 SC
Government Rd 90-D9 J; 90-D9 Y
Governor Berkley Rd 89-C4 W
Governor Dr E 140-D6 NN
Governor Dr W 140-D6 NN
Governor Yeardley La 104-C2 J
Governors Ct 89-D9 W
Governors Dr 89-C9 W
Governors Landing Rd 87-J10; 103-J1 J
Governors Sq 89-B7 J
Grab Bag St 127-F10 Pq
Grace Cir 126-G7 Pq
Grace Dr 140-F2 NN
Grace St 159-E7; HE-C8 H; 125-J1 Y
Grafton Dist Rd 108-K9; 109-A9 Y
Grafton Dr 125-F4 Y
Grafton Sta La 125-F5 Y
Graham Dr 140-G9 NN
Graham Hgts Rd 159-G6; HE-F3 H
Granada Ct 124-H7 NN
Grand Bay Cove 141-E2 NN
Grandview Dr 144-F8 H
Granella St 159-E3 H
Granger Cir 126-A8 Y
Granger Dr 158-F6 H
Granite Pl 90-C6 Y
Grant Allen Ct 142-E7 H
Grant Dr 124-A7 NN
Grant St 159-D6; HE-A4 H
Granville Dr 140-H8 NN
Gravel Hill Rd 58-K9 G
Graves Cir 140-J5 NN
Graves Landing Rd 54-A10; 70-A2 CC
Gray Ave 143-D5 L
Gray Gables Dr 107-B6 J
Gray La 125-H8 Y
Gray St 91-E5 Y
Grayfox Cir 88-H8 J
Grays Deed 105-K2 J
Grays Landing 157-K2 H
Grayson Ave 157-K6 NN
Great Glen 88-K3 J
Great Lakes Dr 159-G2 H
Great Oak Cir 141-E8 NN
Great Pk Dr 124-H6 NN
Great Valley YE-D3 Y
Greate Rd 108-K1 G
Green Ct 157-F3 NN
Green Dr N 141-E4 NN
Green Glen Dr 140-D5 NN
Green Grove La 124-E6 NN
Green Hill La 159-A5 H
Green Long Blvd 141-J2 Y
Green Meadows Dr 124-C10 NN
Green Oaks Rd 157-F3 NN
Green Pl 72-D10; 88-D1 J
Green Spring Ct 159-H2 H
Green St 159-D4 H
Green Swamp Rd 88-C4 J
Green Tree Cove 140-K5 NN
Greenbriar Ave 158-G10; 174-G1 H
Greenbrier 103-K1 J
Greene Dr 125-C2 Y
Greenfield Ave N 158-K2; 159-A2 H
Greenfield Ave S 158-K2; 159-A2 H
Greenfield La 158-K2 H
Greenhouse La 160-A5 H
Greenland Dr 125-J10; 141-J1 Y
Greenlawn Ave 174-F1 H
Greenleigh Ct 107-B7 J
Greens Way 106-F5 J
Greenspeed La 104-H5 J
Greenspring Plantation Dr 88-D9 J
Greensprings Rd 88-B10; 104-D2 J
Greenville Ct 159-K1 H
Greenway Ave 158-A10 NN
Greenway Cir 57-K9 J
Greenwell Dr 142-J10 H
Greenwich La 141-H7 NN
Greenwich Mews 89-B3 J
Greenwood Dr 158-C6, E5 H; 90-C4 Y
Greenwood Rd 157-F3 NN
Greer Pl 124-B10 NN
Gregg La 73-D8 Y
Gregg Ct 140-D1 NN
Gregg Rd 142-K6 H; 143-A6 L
Gregory Ct 143-K10 H
Gregory Dr 126-A1 Y
Gregson Ct 158-B5 H
Grenelefe 88-G6 J
Grenwich La 58-G8 G
Gresham Cir 123-K10 NN
Gretna Ct 144-A10 H
Gretna Way 123-J4 NN
Greyabbey Cir 89-A8 J
Greystone Trace 141-E3 NN
Greystone Walk 125-E4 Y
Griffin Ave 89-G8; WE-B4 W
Griffin St 159-C5 H; 142-F3 Y
Griffith St 160-B9 FM
Grimes Rd 159-K3 H
Grindstone Turn 141-K4 Y
Grissom Way 124-C10 NN
Grist Mill Ct 88-G7 J
Grist Mill Dr 159-A6 H
Grommet La 123-H9 NN
Groome Rd 141-E9 NN
Grossman Pl 158-C8 NN
Grove Ct 124-E6 NN
Grove Hall Cir 141-B1 NN
Grove Hgts Ave 107-A5 J
Grove Rd 107-C5, E5 J
Grove St 160-B4 H; 89-K7; WE-F2 W
Grundland Dr 144-E7 H
Guenevere Ct 141-F5 NN
Guesthouse Ct 104-J4 J
Guinness Rd 124-F6 NN
Gulfstream Dr 124-K9 NN
Gulick Dr 160-B9 FM
Gullane 88-E3 J
Gum Grove Dr 157-G2 NN
Gum Rock Ct 141-D6 NN
Gumwood Dr 158-D5 H
Gunby Rd 141-H10; 157-H1 NN
Gunston Ct 124-H5 NN
Gunston Hall Ct 159-J2 H
Gunter Ct 158-C2 H
Gunter Dr 158-C2 H
Gunter La 158-B2 H
Gurley Ct 160-B3 H
Gurwen Dr 155-C6 I
Guthrie Rd 158-E6 H
Guy La 140-J3 NN
Guy St 159-C5; HE-A1 H
Gwynn Cir 140-E1 NN
Gwynn Rd 140-G9, H9 NN

H

H St 175-H8 NBN
Habersham Dr 140-J1 NN
Hackberry Pl 158-C1 H
Hadley Pl 124-H6 NN
Hadlock Ct 89-F6; WE-B1 Y
Hagood St 123-H7 F
Hague Close 89-F9 W
Hahn Pl 140-D5 NN
Hailsham Pl 124-J6 NN; 125-J4 Y
Hale Dr 159-J3 H
Haley Dr 158-D9 H
Half Penny Dr 55-K5 J
Halifax Ave 159-H6 H
Halifax Pl 141-C2 NN
Hall Rd 144-C10, D8, D10 H
Halles Run 142-E1 Y
Hallmark Dr 140-G10 NN
Hallmark Pl 159-J1 H
Halperin Walk 157-E3 NN
Halsey Rd 107-F4 Y
Halstead La 104-F2 J
Halstead Rd 107-H6 NN; 107-H6, K5 Y
Halyard Dr 124-D3 NN
Hamder Way 125-B10 NN
Hamilton Cir 126-E8 Pq
Hamilton Ct 142-A5 Y
Hamilton Dr 141-A3 NN
Hamilton Rd 109-B5, C5 Y
Hamilton St 174-J1 H; 89-K6; WE-F2 W
Hamlet Ct 141-K5 Y
Hamlet La 159-A4 H
Hamlin Ct 88-F2 J
Hamlin St 157-F4 NN
Hammond Ave 143-D8 L
Hammond St 157-F5 NN
Hampshire Dr 159-D1 H
Hampstead Ct 143-E10 H
Hampton Ave 174-H2 H; 174-D3 NN
Hampton Blvd 175-K10 NBN
Hampton Club Dr 158-F2 H
Hampton Dr 70-F2 J; 174-F3 NN
Hampton Harbor Ave 159-G7; HE-G5 H
Hampton Hwy 141-H1; 142-D4 Y
Hampton Key 106-C2 J
Hampton Roads 158-B1 H
Hampton Roads Ave 159-A8 H
Hampton Roads Bltwy 158-F9; 159-A5, J9; 175-K1; HE-C1 H; 174-E7 NN
Hampton Roads Ctr Pkwy 141-J10 H
Hampton St 159-K10 FM
Hamrick Dr 159-A2 H
Hanbury Pl 124-H5 NN
Hancock Dr 141-A3 NN
Hanes Way 58-J8 G
Hankins Dr 158-H7 H
Hankins Ind Pk Rd 56-B9 J
Hannah St 158-H9 H
Hanover St 159-C9 H
Hanover Way 124-F7 NN
Hansford Ct 110-A8 Y
Hansford La 110-E10 Y
Hansom Dr 126-G7 Pq
Hanson Dr 141-A1 NN
Happiness Pl 124-C6 NN
Happy Acres Rd 143-F10 H
Harbin Ct 104-F1 J
Harbor Cres 110-A8 Y
Harbor Dr 159-A10 H
Harbor Hills Dr 92-J7 G
Harbor La 174-C5 NN
Harbor Rd 103-B1 J; 174-A5, B6, D6 NN
Harbor Watch Pl 140-H8 NN
Harborview La 140-F7 NN
Harbour Dr 109-E7 Y
Harbour Town 88-G5 J
Harbour View Dr 126-G5 Pq
Harbourside 88-J5 J
Harcourt Pl 140-D3 NN
Hard Wood Dr 158-B1 H
Hardee Ct 158-C7 H
Hardt Crossing 123-H2 NN
Hardwick Rd 140-E5 NN
Hardwood Tr 124-E6 NN
Hardy Cash Dr 158-F1 H
Hardy Dr 140-G4 NN
Hardy La 159-E8 H
Hare Rd 158-D8 H
Harlan Dr 125-F3 Y
Harlan Rd 107-F4 Y
Harland Ct 158-D7 H
Harlech 88-G3 J
Harlech Pl 124-J6 NN
Harlow Ct 143-K9 H
Harness Ct 39-G10 J
Harpers Dr 141-J7 NN
Harpers La 158-D8 H
Harpers Mill 87-E10 J
Harpersville Rd 141-G5, G10; 157-E2 NN
Harpoon Ct 107-B7 J
Harrad La 72-F3 J
Harrell Ct 124-E10 NN
Harrell La 92-G3 G
Harrells Ct 106-A3 J
Harriett St 159-C8; HE-A8 H
Harrigan Way 141-F2 Y
Harrington Rd 140-D3 NN
Harris Ave 159-G8; HE-G8 H; 143-D6 L
Harris Cr Rd 143-J10, K6; 159-H1 H
Harris Grove La 109-A9 Y
Harris La 109-C9 Y
Harris Landing Rd 143-K9 H
Harris Rd 140-G8 NN
Harrison Ave 89-F6; WE-B2 W
Harrison Ct 142-B5 Y
Harrison Loop 123-G6 F
Harrison Rd 123-A8, C6 F; 157-F5 NN
Harrison St 160-A10 FM; 159-E5; HE-C3 H
Harrod La 109-B9 Y
Harrogate La 142-C10 H
Harrop La 90-A8 Y
Harrop Parrish Ct 88-F2 J
Harrops Glen 106-E3 J
Harry Ct 158-F1 H; 158-C8 NN
Harston Ct 124-E10 NN
Hart Cir 159-K3 H
Hartford Rd 158-G3 H
Hartless Ct 160-B1 H
Harvest Cir 104-K2 J
Harvest La 160-D1 H
Harvest Way 141-J5 Y
Harvey St HE-B2 H
Harwin Dr 158-B4 H
Harwood Ct 106-A4 J
Harwood Dr 160-B4 H; 123-F2 NN; 125-H8 Y
Harwood La 107-E4 J; 107-E4 Y
Hastings Dr 159-J3 H
Hastings La 88-J1 J
Hatcher Ct 142-E4 Y
Hatchland Pl 124-J5 NN
Hatfield Pl 124-G5 NN
Hatteras Landing 143-F9 H
Hatton Cir 142-H10 H
Haughton Ave 141-A7 NN
Haughton La 140-G9 NN
Haven Ct 140-D2 NN
Haverstraw Ct 125-G6 Y
Haverty La 155-C8 I
Haviland Dr 157-H2 NN
Havlow La 155-C7 I
Havoc Dr 142-G8 Y
Hawk Ct 142-B5 Y
Hawk Nest Ct 142-A6 H
Hawkeye Ct 124-C4 NN
Hawkins Ave 92-J5 G
Hawkins Ct 142-C6 H
Hawley Dr 92-H3 G
Hawthorn Pl 158-D1 H
Hawthorne Pt 125-B1 Y
Hawthrone Dr 140-F2 NN
Hayes Dr 140-F4 NN
Hayes Rd 92-K5 G
Hayes Shopping Ct 92-H3 G
Haymaker Pl 90-F10 Y
Haynes Dr 89-K5; WE-E1 W
Haystack Landing Rd 140-A3 NN
Haywagon Tr 143-J8 H
Hazelwood Ave 56-C6 J
Hazelwood Ct 124-G5 NN
Hazelwood Rd 157-K6; 158-A6 H
Heacox La 124-G6 NN
Headquarters Way 141-A4 NN
Headrow Terr 158-B3 H
Healey Dr 123-K9 NN
Hearthside La 104-J2 J
Hearthstone 125-E4 Y
Hearthstone Way 124-G6 NN
Heartwood Crossing 55-K2 J
Heath Dr 92-F2 G
Heath Pl 141-B8 NN; 141-H2 Y
Heather Cir 158-K7 H
Heather Ct 127-D10 Pq
Heather La 73-B9 J; 140-J6 NN
Heather Way 141-J2 Y
Heatherwood La 125-F2 Y
Heathery 88-J3 J
Heathland Dr 141-D3 NN
Heaths La 72-H6 J
Heavens Way 142-B1 Y
Hedgelawn Ct 159-H9 H
Hedgerow La 125-F10; 141-F1 Y
Heffelfinger St 159-C5, D5; HE-A1, A2 H
Heiner Pl 123-H5 F
Helen Dr 140-A1 NN
Helena Dr 124-B9 NN
Helm Ave 143-C5 L
Helm Dr 140-C2 NN
Helmsly Rd 89-E10 W
Hemlock Ct 141-J4 Y
Hemlock Rd 157-G3 NN
Hemlock St 158-G9 H
Hempstead Ct 88-J1 J
Hempstead Rd 88-H2 J
Henderson Dr 159-K5 H
Henderson St 89-G5 W
Henley Dr 89-B4 J
Henrico Ct 124-H5 NN
Henry Clay Rd 157-G2 NN
Henry Ct 142-B4 Y
Henry Hogge La 92-F3 G
Henry Lee Dr 107-J6 NN; 107-J6 Y
Henry Lee La 125-J5 Y
Henry St 159-D8 H; 105-H1 W
Henry St N 89-G6, H7; WE-C2, C3 W
Henry St S 89-G9; WE-C4 W
Henry Tyler Dr 88-F2 J
Henry Whitfield St 127-B9 Pq

Hensley Ct 140-G4 NN
Hensley Dr 140-G4 NN
Herbert Ave 160-C4 H
Hercules Ave 142-B4 Y
Hercules Dr 159-G1 H
Heritage Ests La 92-G2 G
Heritage Landing Rd 88-A10; 104-A1 J
Heritage Pl 125-H2 Y
Heritage Pointe 88-J5 J
Heritage Way 140-G1 NN
Hermitage Pl 160-B5 H
Hermitage Rd 140-J8 NN; 89-A9 W
Herndon Jenkins Rd 73-A7 J
Heron Cir 143-B1 Pq
Heron Ct 72-K10 J; 109-C8 Y
Herring Pl 141-J7 H
Herstad Ct 72-F3 J
Hertzler Rd 140-A3 NN
Hess St 159-D8; HE-B8 H
Hewins La 153-K9 H
Hiawatha Ct 104-H4 J
Hickory Ave 174-G1 NN
Hickory Ct 88-A6 J
Hickory Fork Rd 58-J3 G
Hickory Hill Dr 158-D5 H
Hickory Hills Dr 90-D7 Y
Hickory La 88-A6 J; 142-C4 Y
Hickory Pt Blvd 124-A6 NN
Hickory Rd 143-C8 L
Hickory Signpost Rd 104-J1 J
Hickory St 159-E2 H
Hicks Is Rd 54-G7 J
Hidalgo Ct 143-J10 H
Hidalgo Dr 143-J10 H
Hidden Dr 38-A6 NK
Hidden Glen Ct 124-D8 NN
Hidden Harbor 126-B2 Y
Hidden La 38-A6 NK; 110-C8 Y
Hidden Lake Dr 105-C1 J
Hidden Lake Pl 140-G4 NN
Hidden Nest 56-A1 J
Hide-Away La 73-F1 J
Hiden Blvd 140-J10; 141-A10 NN
Hidenwood Dr 140-K9; 141-A8, A9 NN
Higgins La 160-D1 H
Higginson Ct 88-G3 J
High Bluff La 39-K1 NK
High Ct La 159-E6; HE-D4 H
High Dunes Quay 160-E2 H
High Pt Rd 90-F9 Y
Highbank La 70-D3 CC
Highfield Dr 55-K5 J
Highgate Green 88-G1 J
Highland 88-D1 J
Highland Ave 159-B8 H
Highland Ct 158-A9 NN
Highland Dr 92-K8 G
Highlands Pkwy 107-H9 NN
Highwood Ctr 123-K9 NN
Hiking Tr 90-G4 Y
Hilda Cir 158-D8 H
Hilkey Dr 142-J7 H
Hill Rd 92-K8 G
Hill St 174-J1 H
Hillard Cir 140-J4 NN
Hillburne La 125-G6 Y
Hillcrest Cir 158-K3 H
Hillcrest Dr 140-G10 NN
Hilliard St 143-D5 L
Hillsboro Dr 157-K5 H
Hillside Dr 158-E6 H
Hillside La 90-C6 Y
Hilltop Ct 55-F7 J
Hilltop Dr 123-J1 NN
Hilmar Pl 158-C8 NN
Hilton Blvd 158-A9 NN
Hilton Pkwy 157-F5 NN
Hilton Terr 157-F6 NN
Hines Ave 174-A1 NN
Hines Cir 123-H5 F
Hinton Dr 159-J4 H
Hinton St 72-B1 J
Hiram Crockett Ct 143-F10 H
Historical Tour Rd 108-D4, E6, G8, H6, J6; 109-A4, A7; YE-C6 Y
Hitchens La 157-D2 NN
Hobson Ave 174-G1 H
Hockaday Rd 54-H7, J7; 55-A7 J
Hodges Cove Rd 126-C4 Y
Hodges Dr 159-A3 H
Hodges La 140-K7 NN
Hoefork La 92-K8 G
Hofstadter Rd 141-B5 NN
Hog Is Rd 106-A10 SC
Hogan Dr 141-A4, A5 NN
Hogges La 92-G3 G
Hogster La 92-G2 G
Holbrook Dr 141-A1, B1 NN
Holcomb Dr 90-C4 Y
Holden Ct 140-G9 NN
Holden La 125-H4 Y
Holdsworth Rd 106-C2 J
Holiday Dr 143-K9 H
Holland Ct 124-J6 NN
Hollingsworth Ct 125-E10 Y
Hollins Ct 124-E6 NN
Hollinwell 88-D1 J
Hollis Wood Dr 158-B1 H
Holliston Blvd 141-A1 NN
Hollomon Dr 158-C5 H
Hollow Cr Ct 143-F9 H
Hollow Oak Dr 56-A1 J
Hollow Pond Rd 92-G1 G
Holloway Dr 157-K1 H; 90-D5 Y
Holloway Rd 140-C5 NN; 127-A9 Pq
Holly Bk Dr 104-C1 J
Holly Dr 157-D4 NN
Holly Forks Rd 39-G10 J; 40-A3, D6, D7 NK
Holly Grove 104-J2 J
Holly Hills Dr 89-D9 W
Holly La 54-G10 J
Holly Pt Rd 86-J9 CC; 126-C5 Y
Holly Rd 105-A2 J; 143-C8 L
Holly Ridge La 104-J2 J
Holly St 159-G5; HE-G2 H; 143-B9 L; 143-C1 Pq
Hollyberry St 159-C10 H
Hollywood Ave 174-H1 H
Hollywood Blvd 125-J8 Y
Holm Rd 91-J9 Y
Holman Rd 106-C1 J
Holmes Blvd 125-F3 Y
Holston La 144-C8 H
Holt Cir 125-K1 Y
Home Pl 159-H8 H
Homeland St 174-G1 H
Homestead Ave 159-B7 H
Homestead Pl 142-A1 Y
Homestead Rd 38-G10, J8; 39-A6 NK
Hondo Ct 143-J10 H
Honeysuckle Hill 143-K9 H
Honeysuckle La 123-H9 NN; 141-J5 Y
Hook Rd 92-K6 G
Hoopes Rd 124-D10 NN
Hop Ct 124-F10 NN
Hope Ct 140-H3 NN
Hope St 159-H8 H
Hopemont Cir 159-H9 H
Hopemont Dr 140-F10 NN
Hopkins Ct 142-H1 Pq
Hopkins St 159-J7 H; 157-G6 NN
Horan Ct 73-C10 J
Horizon La 124-H8 NN
Horn Hawthorn La 105-C2 J
Hornes Lake Dr 87-D10, E10; 103-E1 J
Hornet Cir 157-H8 NN
Hornet St 158-G9 H
Hornsby La 140-K2 NN
Hornsbyville Rd 109-D9, E10, F9 Y
Horse Pen Rd 140-D7 NN
Horse Run Glenn 140-F1 NN
Horse Shoe Dr 90-E3, E4 Y
Horse Shoe Rd 70-A3 CC
Horseshoe Ct 124-E9 NN
Horseshoe Landing 159-H1 H
Horsley Dr 158-D4 H
Horton Rd 144-C8 H
Hosier St 157-G5, G6 NN
Hounds Chase 125-F10 Y
House Dr 141-D7 NN
House of Burgesses Way 89-G4 Y
Houston Ave 159-K1 H
Howard Ave 175-H9 NBN
Howard Ct 157-F1 NN
Howard Dr 106-K4; 107-A4 J
Howard St 159-J8 H
Howe Rd 143-J8 H
Howmet Dr 158-G8 H
Hoylake 88-D3 J
Hubbard La 90-C8 Y
Hubbard Rd 54-G3 NK
Huber Rd 141-E10, G9 NN
Hudgins Farm Dr 125-G4 Y
Hudgins Rd 92-J5 G; 126-H10 Pq
Hudson Cir 157-H5 NN
Hudson Dr 73-F7 Y
Hudson Pl 157-H4 NN
Hudson Terr 157-H5 NN
Huffman Dr 159-E4, E5 H
Hughes Ct 157-E1 NN
Hughes Dr 175-J7, K8 NBN; 124-C9 NN
Hughes La 160-A6 H
Huguenot Rd 141-C9 NN
Hull Dr 158-D8 H
Hull St 175-H8 NBN; 157-E2 NN
Hundley Dr 141-H3 Y
Hunlac Ave 160-C2 H
Hunsaker Loop 142-J4 H
Hunstanton 88-J4 J
Hunt Blue Run Cir 158-D3 H
Hunt Club Blvd 158-E3 H
Hunt Club Run 123-J4 NN
Hunt Wood Dr 142-G2 Pq
Hunter La 90-E4; 126-B2 Y
Hunter Rd 157-F5 NN
Hunter Tr 143-J9 H
Huntercombe 88-C3 J
Hunters Glen 140-J5 NN
Hunters Ridge 88-G9 J
Huntgate Cir 140-K8 NN
Hunting Ave 143-E8, F7 L
Hunting Cove 89-G9 W
Hunting Towers Ct 107-B5 J
Huntingdon Rd 90-C5 Y
Huntington Ave 157-J10; 173-K1; 174-A3 NN
Huntington Dr 88-K9 J
Huntington Rd 159-F7; HE-F8 H
Hunts Neck Rd 126-F6 Pq
Huntshire La 126-K10; 127-A10 Pq
Huntstree Pl 140-H4 NN
Hurley Ave 157-F5 NN
Hurst Dr 159-J3 H
Hurst St 89-B10 J
Huskie Dr 142-B4 Y
Hustings La 124-D9 NN
Huxley Pl 141-B7 NN
Hyacinth Cir 124-C10 NN
Hyatt Pl 140-J5 NN
Hyde La 141-H2 Y
Hygeia Ave 159-K6 H

I

Ibis Pl 126-B4 Y
Ida St 159-B7 H
Idaho Cir 73-F4 Y
Idlewood Cir 140-J9 NN; 125-H10 Y
Idlewood Dr 89-G8 W
Ignoble La 110-C6 Y
Ilene Dr 124-B9 NN
Ilex Ct 158-J2 H
Ilex Dr 126-A4 Y
Impala Dr 141-E8 NN
Incinerator La 160-C7 FM
Incubate Rd 159-G2 H
Independence Dr N 158-J6 H
Independence Dr S 158-J6 H
Independence Pl 141-B4 NN
Indian Field Rd 92-A10 Y
Indian Cir 107-B6 J
Indian Path 88-G9 J
Indian Rd 159-F7; HE-E7 H
Indian Springs Ct 89-F8; WE-B4 W
Indian Springs Dr 140-J7 NN
Indian Springs Rd 89-F8; WE-B4 W
Indian Summer Dr 142-E5 Y
Indian Summer La 73-B8 J
Indiana La 73-E4 Y
Indigo Dam Rd 88-K7; 89-A8 J; 156-J1 NN
Indigo Terr 89-A7 J
Industrial Blvd 56-B8 J
Industrial Pk Dr 124-B4, C4, D4 NN
Industry Dr 158-G8 H; 141-H1 Y
Ingalls Rd 159-K8 FM
Ingham St 109-D6 Y
Inglewood Dr 158-K3; 159-A3 H
Ingram Rd 88-G10 J
Inland View Dr 123-J1 NN
Inlandview Dr 158-J7 H
Insley Cir 126-F8 Pq
Institute Whipple Dr 159-K5 H
Invader Dr 142-C8 Y
Ira Ct 158-F6 H
Ira Dr 125-J2 Y
Ireland St 159-J6, K5 H; 89-G8; WE-C4 W
Iris La 123-K6 NN
Iris Pl 160-A6 H
Ironbound Rd 88-H10, J7, K6; 89-A7, B6; 104-J1 J; 89-B6, C4 W
Ironbridge Ct 160-A2 H
Irongate Ct 141-B1 NN
Ironmonger La 110-C8 Y
Ironwood Dr 89-B2 W; 141-F1 Y
Irwin St 123-H7 F
Isaac Ct 124-E2 NN
Isaacson Rd 91-C5 Y
Isham Pl 124-H6 NN
Island Cove Dr 159-F5 H
Island Ct 127-A8 Pq
Island Quay 141-E2 NN; 141-E2 Y
Island Rd 58-D2, F1 G
Island Rd E 103-G1 J
Island Rd W 103-F1 J
Island View Dr 140-D6 NN
Islander Way 126-H10 Pq
Isle of Wight Ct 104-K1 J
Iverness 88-J4 J
Ivory Gulf Cres 160-E2 H
Ivy Arch 141-J4 Y
Ivy Ave 174-D4 NN
Ivy Ct 105-H1 J
Ivy Farms Rd 157-G2 NN
Ivy Hill Rd 55-E8 J
Ivy Home Rd 159-C8, D8 H
Ivystone Way 141-E3 NN

J

J Clyde Morris Blvd 141-C10, G8 NN
J Farm La 73-C4 Y
Jacinth Cir 124-C10 NN
Jack Shaver Dr 123-J4 NN
Jacklyn Cir 157-K1 H
Jacks Pl 124-C4 NN
Jackson Ave 123-H6 F
Jackson Cir 141-K1 Y
Jackson Ct 142-A4 Y
Jackson Dr 90-C8 J
Jackson Rd 157-H4 NN
Jackson St 106-K5; 107-A4 J
Jacobs La 140-K10; 156-J1 NN
Jacobs Rd 90-D10 J
Jacobs Run 125-H5 Y
Jacqueline Dr 92-J6 G
Jakes La 124-D4 NN
James Blair Dr 89-F7; WE-B3 W
James Bray Dr 88-F1 J
James Ct 159-G8; HE-F8 H
James Dr 92-J7 G; 54-A7 NK; 158-B9 NN; 125-G4 Y
James Landing Rd 156-H1 NN
James Longstreet 106-B2 J
James River Dr 157-D5 NN
James River La 140-F10 NN
James Sq 105-C1 J
James Terr 158-D4 H
Jameson Ave 159-B2 H
Jamestown Ave 158-J9 H
Jamestown Dr 140-K8 NN
Jamestown Rd 89-C10; 104-G3; 105-A2 J; 89-C10, E8; WE-B4 W
Jameswood 104-A1 J
Jamie Ct 124-A10 NN
Jan La 110-D10 Y
Jan Rae Cir 107-C7 J
Jan-Mar Dr 141-B10 NN
Janet Cir 141-D10 NN
Janet Dr 158-D8 H
Janice Ct 157-J2 H
Janice Rd 40-B8 NK
Janis Dr 109-F7 Y
Jarvis Pl 158-A7 NN
Jasmine Cir 124-D10 NN
Jasmine Ct 160-A6 H
Jason Dr 90-E8 Y
Jay Sykes Ct 160-A4 H
Jaymoore La 159-J5 H
Jayne Lee Dr 160-C2 H
Jean Mar Dr 126-J10; 142-J1 Pq
Jean Pl 126-A10 Y
Jebs Pl 174-F4 NN
Jefferson Ave 123-J5 F; 107-J8; 108-A10; 124-B1, E6, H9; 140-J1; 141-A3, D7; 157-F1, J7; 174-B1 NN
Jefferson Ct 126-G8 Pq; 142-A4 Y
Jefferson Hundred 106-E4, F4 J
Jefferson La 125-F2 Y
Jefferson Pt La 141-A4 NN
Jefferson St 89-K6; WE-E2 W
Jeffery La 158-D8 H
Jefferys Dr 157-G2 NN
Jeffrey Kenneth Pl 142-E1 Y
Jenkins Ct 141-F1 Y
Jenkins La 144-C8 H
Jenness La 124-G10; 140-F1 NN
Jennie Dr 108-K8 Y
Jennifer La 159-F4 H
Jennifer Pl 124-A8 NN
Jennings Dr 126-E3 Y
Jennisons Fall 159-F2 H
Jerdone Pl 109-B9 Y
Jerdone Rd 106-C1 J
Jericho Crossing 125-H5 Y
Jernigan La 126-C4 Y
Jerome Cir 159-J4 H
Jesse St 141-F1 Y
Jessica Dr 142-E1 Y
Jessie Cir 124-E2 NN
Jester Ct 124-E5 NN
Jesters La 88-H6 J
Jethro La 125-K3 Y
Jib Ct 160-D1 H
Jimmy Ct 158-B1 H
Joan Cir 141-E10 NN
Joan Dr 158-K2 H
Joanna Pl 140-H5 NN
Joanne Cir 126-A1 Y
Joanne Ct 56-C6 J
Joanne Dr 110-A10; 126-A1 Y
Jockeys Neck Tr 105-D4, E3 J
Joel Ct 124-D6 NN
Joel La 125-H7 Y
John Bratton 105-K4 J
John Browning 106-F5 J
John Fowler 106-B3 J
John Jefferson Rd 106-E2 J
John Paine 106-B3 J
John Pinckney La 89-C3 W
John Pot Dr 88-G1, G3 J
John Proctor E 104-H2 J
John Proctor W 104-G1 J
John Ratcliffe 106-B2 J
John Rolfe Dr 124-F9 NN
John Rolfe La 88-A10; 104-A2 J
John Rolfe Rd 92-F4 G
John Shropshire 88-B7 J
John Smith La 92-F3, F4 G
John Smith Tr 54-E9 NK
John Twine 106-A4 J
John Tyler Mem Hwy 86-A4, F7 CC; 87-B8, G10; 88-B10, H10; 89-A10, B9; 104-G1 J; 89-B9 W
John Vaughn Rd 87-J10; 103-J1 J
John Wickam 106-E4 J
John Wythe Pl 89-C3 W
Johns Landing Cir 126-G6 Pq
Johnson Dr 144-C7 H
Johnson La 124-C8 NN
Jolama Dr 140-A1 NN
Jolly Pond Rd 71-H9; 72-C7; 87-G1, J3 J
Jonadab Dr 125-K8 Y
Jonadab Rd 125-K8; 126-A8 Y
Jonas Profit Tr 104-C2 J
Jonathan Ct 124-C9 NN
Jonathan Jct 141-H2 Y
Jonathon Ct 72-C1 J
Jones Dr 72-E7 J; 90-J7 Y
Jones Mill La 89-D9 W
Jones Mill Rd 88-G7 J
Jones Rd 159-D8 H; 157-D3 NN
Jonestown Rd 41-J1 KQ
Jonquil Ct 142-A2 Y
Jonquil La 140-G7 NN
Jordan Dr 158-F5 H; 140-H10 NN
Jordans Journey 104-B2, C2 J
Joseph Lewis Rd 92-J8 G
Joseph Topping Rd 142-K2 Pq
Josephs Crossing 159-D1 H
Joshua La 159-K2 H
Joshua Way N 125-G5 Y
Joshua Way S 125-H5 Y
Jotank Turn 142-C3 Y
Jouett Dr 124-G7 NN
Joy Dr 158-A5 H
Joyce Cir 141-E10 NN
Joyce Lee Cir N 158-F6 H
Joyce Lee Cir S 158-F6 H
Joynes Rd 158-D7 H
Joys Cir 72-B1 J
Juanita Dr 158-D7 H
Jubal Pl 106-B2 J
Judges Ct 124-E8 NN
Judith Cir 141-K5 Y
Judy Ct 159-B7 H
Judy Dr 107-B7 J; 124-A5 NN
Jules Cir 141-G10 NN
Julia Terr 124-F6 NN
Julian Pl 158-J3 H
June Terr 107-H9 NN
Juniper Ct 107-A5 J
Juniper Dr 157-H3 NN
Juniper Hills 88-G5 J
Juniper St 159-G5; HE-H2 H
Jurasin Pl 123-J5 F
Jury La 124-E8 NN
Justice Grice 106-A2 J
Justice La 73-G5 J
Justin La 157-F7 NN
Juvnal Rd 107-G5 Y

K

K Ave 109-G8 Y
Kaitlyn Cir 142-E7 H
Kaitlyn Ct 141-H4 Y
Kaleigh Ct 157-J2 H
Kanawah Run 142-C2 Y
Kanawha Cir 157-H7 NN
Kanawha Ct 143-H10 H
Kansas Ct 160-B2 H
Karen Dr 124-A7 NN
Kass Ct 124-C9 NN
Kathann Dr 157-J5 H
Katherine Ct 141-G6 NN
Katherine St 159-B8 H
Kathleen Pl 92-K7 G; 141-K5 Y
Kathleen Way 88-K7 J
Kathryn Ct 89-C10 J
Kathy Dr 126-K10 Pq
Katies Ct 140-F10 NN
Katspaw Ct 124-E5 NN
Kay Cir 126-B10 Y
Kay La 141-K3 Y
Kayla Ct 142-E1 Y
Kearny Ct 124-A8 NN
Kecoughtan Rd 158-K10; 159-A9, D8; 174-G1 H
Keel Ct 158-F4 H; 124-D3 NN
Keepers Hill 103-B1 J
Keeton Ct 158-E3 H
Keith Pt 174-F4 NN
Keith Rd 159-D2 H; 140-H6 NN
Kells Dr 123-E5 F
Kelly Ave 159-J6 H
Kelly Pl 140-G5 NN
Kelsey Rd 89-F4 Y
Kelso Dr 141-H8 NN
Kelsor Rd 126-G9 Pq
Kemp La 92-J4 G
Kempe Dr 88-G1 J
Kemper Ave 141-D10 NN
Kemper Charge 143-H10 H
Kempton St 159-E5; HE-C1 H
Kendale La 88-K2 J
Kendall Dr 157-G2 NN
Kenilworth Dr 157-K2 H; 140-K8 NN
Kenmar Dr 125-J4 Y
Kenmore Dr 159-A10 H
Kennedy Dr 157-K6 H
Kenneth Dr 110-B9 Y
Kenneth St 159-C7 H
Kensington Ct 88-H1 J
Kensington Dr 159-K4 H; 140-E4 NN
Kensington Pl 141-K4 Y
Kent St 174-J1 H
Kent Taylor Dr 142-A3 Y
Kentucky Ave 158-H9 H
Kentucky Dr 125-A7 Y
Kentwell Ct 124-H6 NN
Kenwood Dr 158-K3 H
Keppel Dr 124-G7 NN
Kerlin Rd 141-F10 NN
Kerr Rd 123-D4 F
Kerry Lake Dr 140-A3 NN
Kersten 88-D2 J
Kestral Ct 72-K9 J
Kestrel Ct 140-G9 NN
Keswick Cir 140-C1 NN
Keswick La 158-J7 H
Ketch Ct 124-E3 NN
Kettering La 142-C10 H
Kevin Ct 140-D1 NN
Key Cir 125-J9 Y
Key Oak Dr 126-K10 Pq
Key Stone 88-J4 J
Keyser La 92-G2 G
Keystone Dr 142-C7 Y
Kicotan Turn 142-D1 Y
Kill Deer La 142-A6 H
Killarnock Ct 141-D1 Y
Killington 88-J5 J
Kiln Cr Blvd 141-C1 NN
Kiln Cr Pkwy 141-C2 NN; 125-D10; 141-C2, E1 Y
Kimberly Ave 159-H6, J6 H
Kimberly Ct 140-G4 NN; 125-G8 Y
Kincaid La 141-J8 H
King Arthur Ct 124-E5 NN
King Cobra Ct 142-C8 Y
King Ct 90-A9 J
King Forest La 123-J9 NN
King Henry Way 73-B10; 89-B1 J
King Kove La 159-D1 H
King Rd 107-E4 Y
King Richard Ct 90-C4 Y
King Richard Pl 140-E4 NN
King St HE-C1, D3 H
King St E 143-E10 H
King St N 159-E3, E5 H
King St S 159-E7; HE-D5 H
King William Dr 73-B10 J
Kings Cr La 141-B1 NN
Kings Ct 140-G10 NN
Kings Grant Ct 123-J3 NN
Kings Grant Dr 126-C5 Y
Kings Landing La 159-H4 H
Kings Pointe Crossing 141-H4 Y
Kings Pt Dr 159-G4 H
Kings Ridge Dr 124-E5 NN
Kings View Ct 160-A1 H
Kings Way 159-E6; HE-D4 H; 89-B9 J
Kingsbridge La 125-G6 Y
Kingsbury Dr 140-K7 NN
Kingsgate Rd 73-E3 Y
Kingslee La 159-E1 H
Kingsman Dr 124-B9 NN
Kingsmen Way 158-A6 H
Kingsmill Dr 157-E3 NN
Kingsmill Rd 106-D4 J
Kingspoint Dr 105-G2, H2 J
Kingston Ct 141-K5 Y
Kingston Hall Ct 159-K2 H
Kingstowne Dr 141-C8 NN
Kingstowne Rd 141-C7 NN
Kingsway Ct 159-G3 H
Kingsway Dr 141-C7 NN
Kingswood Dr 89-B10; 105-B1 J
Kingswood Dr W 105-A1 J
Kingwood Dr 141-K7 NN
Kinsale Cres 141-D2 NN
Kinsey Dr 92-H5 G
Kirby Ct 126-J10 Pq
Kirby La 159-C9 H; 142-E1 Y
Kirbys Farm Rd 40-E6 NK
Kirk Cir 123-H5 F
Kirk Dr 124-F7 NN
Kirkland Ct 90-B8 J
Kirkwood Cir 157-K3 H
Kiskiac Turn 142-C2 Y
Kiskiack Rd 92-B8 Y
Kitchems Close 103-H1 J
Kitchums Pond Rd 103-H1 J
Kitty Dr 125-J8 Y
Kittywake Dr 140-B2 NN
Kiwanis Dr 103-A10 SC
Kiwanis St 158-G9 H
Klich Dr 142-J7 H
Knickerbocker Cir 158-K5 H
Knight Ct 123-H2 NN
Knight St 158-H7 H
Knodishall Way 144-E9 H
Knoll Bay La 58-F6 G
Knoll Crest 125-E4 Y
Knollwood Dr 56-E9 J; 123-J10 NN
Knott Pl 89-A8 J
Knox Pl 159-E4 H
Knox St 159-G8 H
Kohl Rd 90-J5, K4; 91-A3 Y
Kohler Cres 141-B8 NN
Kopek Ct 158-F1 H
Kostel Ct 159-G1 H
Kove Dr 159-D2 H
Kraft Ct 141-F1 Y
Kramer Ct 160-C1 H
Krause Ct 144-C10 H
Kristiansand Dr 72-G3 J
Kristos Ct 107-B5 J
Kristy Ct 140-J3 NN
Kroken Ct 72-F3 J
Kubesh Ct 142-F2 Y
Kyle Cir 141-K2 Y
Kyle St 142-G7 H

L

L Ave 109-G8 Y
La Grange Pkwy 56-A5 J
Lackey Dr 141-A7 NN
Lackey Rd 108-C2 Y
Lacon Dr 123-K8 NN
Lacrosse St 159-J6 H
Lacy Cove La 124-D10 NN
Ladd St 159-F7; HE-E6 H
Lady Slipper Path 88-K10 J
Lafayette Blvd 72-K10; 88-J1 J
Lafayette Dr 160-C1, C2 H; 107-H7 NN
Lafayette Hgts Dr 92-K10 G
Lafayette Rd 126-H8 Pq; 109-B5; YE-G5 Y
Lafayette St 89-F6, J7; WE-A1, C2, E2 W
Lafayette Terr 157-H5 NN
Lafiete La 107-B5 J
Lafluer La 123-H2 NN
Lagoon La 107-B10 J
Laguard Dr 158-J10 H
Lair Cir 144-B10 H
Lake Cir S 141-K7 H
Lake Dale Way 141-K3 Y
Lake Dr 104-K2; 105-A2 J; 141-C1 NN
Lake Erie Ct 159-G2 H
Lake Ferguson Ct 159-G3 H
Lake Field Crossing 141-K7 H
Lake Forest Dr 88-G8 J; 141-C1 NN
Lake Herrin Ct 142-A2 Y
Lake Huron Ct 159-G2 H
Lake Loop N 141-J7 H
Lake One Dr 158-E3 H
Lake Ontario Ct 159-G3 H
Lake Ovide Ct 159-G2 H
Lake Philip Dr 159-G2 H
Lake Pointe Dr 107-G8 NN
Lake Powell Rd 89-C10; 104-K3; 105-A3, B3, D3, J; 89-C10 W
Lake Pt Ct 56-F9 J
Lake Pt Tr 54-F8 NK
Lake Shore Dr E 158-F4 H
Lake Superior Ct 159-G3 H
Lake View Dr 158-E3 H
Lake Walk Crossing 141-K7 H
Lake Way 143-H9 H
Lakecrest Ct 141-C2 NN
Lakeland Cres 142-A3 Y
Lakeland Cres N 141-K1 Y
Lakeland Dr 159-H3 H; 158-C8 NN
Lakepoint Pl 125-J5 Y
Lakeshead Dr 90-C6 Y
Lakeshore Dr 124-A7 NN
Lakeside Cir 105-A3 J
Lakeside Cres 158-J6 H
Lakeside Dr 88-G9 J; 157-B2 NN; 125-G7, K4 Y
Lakeview Dr 71-F1 J; 140-C4 NN; 143-B2 Pq; 126-A7 Y
Lakewood Cir 125-G7 Y
Lakewood Dr 159-A3 H; 104-J2 J
Lakewood Pk Dr 140-C4 NN
Lambert Dr 140-E4 NN

STREETS

Lambs Cr Dr 126-C9 Y
Lambs Rest La 126-D7 Y
Lamington Rd E 159-E1 H
Lamington Rd W 159-D1 H
Lamphier La 140-G10 NN
Lamplighter Pl 106-F1 Y
Lampros Ct 158-D2 H
Lancaster Ct 104-C1 J
Lancaster La 88-H10 J; 141-B2 NN
Lancaster Terr 158-B3 H
Lance Dr 141-F8 NN
Lancelot Ct 140-D5 NN
Lancer St 159-J8 H
Land Grant Rd 110-B7 Y
Landing Cir 57-B7 J
Landing Dr 122-A8 SC
Landing E 106-B4 J
Landing La 140-C5 NN
Landing Rd 104-J3 J; 110-A9 Y
Landing W 106-A5 J
Landmark Ct 124-F7 NN
Landmark Dr 157-K2 H
Landridge Ct 124-G7 NN
Landrum Dr 89-F7; WE-A4 W
Lands End Cir 159-C2 H
Lands End Dr 104-H4 J
Lands End La 123-K6 NN
Landsdown 88-F4 J
Lanelle Pl 124-E3 NN
Langholm Ct 159-K1 H
Langhorne Rd 140-J10; 156-H1 NN
Langille Dr 160-A5 H
Langley Ave 159-C5, D5; HE-A1, A2 H; 141-E10; 157-E1 NN
Langley Blvd 142-H5, J7 H; 143-D9 L
Langley St 143-D1 Pq
Langman Pl 104-J5 J
Langston Blvd 158-F5 H
Langston Pt 126-D7 Y
Lankford La 142-J1 Pq
Lansdown Cir 157-J3 H
Lansing Dr 160-A4 H
Lantana La 159-D3 H
Lantern Cir 156-H1 NN
Lantern Pl 90-F10 Y
Lanyard Rd 140-B2 NN
Larabee La 159-J8 H
Laramie Ct 124-A8 NN
Larchmont Cres 141-C10 NN
Larchmont Rd 92-K7 G
Larchwood Ct 124-K1 Y
Larchwood Rd 108-K10 Y
Laredo Ct 143-K10 H
Larissa Dr 141-H8 NN
Lark Cir 141-G10 NN
Larkin Run 125-H7 Y
Larkspar Run 105-F1 J
Larkspur Hollow 125-E4 Y
Larson Ct 140-A1 NN
Las Brisas Ct 107-A5 J
Lasalle Ave 143-B10 H
LaSalle Ave 159-B2, B4, C8 H
Lassiter Dr 158-C7 H; 174-E5 NN
Lateen Ct 124-D3 NN
Latham Dr 157-G4, H3 NN
Latta La 142-A4 Y
Lauderdale Ave 159-B8 H
Laura La 125-J7 Y
Laurel Acres 109-F8 Y
Laurel Ct 105-C1 J; 157-H4 NN
Laurel Dr 159-F4 H
Laurel Haven Dr 92-F2 G
Laurel La 54-G10 J; 90-E3 Y
Laurel Path Rd 108-K10; 109-A10 Y
Laurel Ridge 39-K10; 55-K1 J
Laurel Wood Rd 140-B2 NN
Lauren Dr 142-H2 Pq
Laurent Cir 123-K8 NN
Lavelle La 104-H4 J
Lavender Trace 160-A2 H
Lawndale Dr 158-G7 H
Lawnes Cir 104-A1 J
Lawnes Cr Rd 103-H1 J
Lawrence Ave 160-A5 H
Lawrence Dr 141-B5 NN
Lawson Dr 125-J10 Y
Lawson Rd N 127-B8 Pq
Lawson Rd S 127-D10 Pq
Laydon Cir 157-A1 NN
Laydon Way 127-B10; 143-B1 Pq
Le Roy Dr 126-A8 Y
Leader La 141-B5 NN
Leafwood La 88-G8 J
Leahs Trace 157-J4 H
Leanne Ct 125-A1 Y
Lear Cir 140-K4 NN
Lear Dr 124-K9 NN
Leatherleaf Dr 55-K1 J
Lebanon Ch Rd 107-G6 NN; 107-G6 Y
Lee Ave 126-G7 Pq
Lee Blvd 123-D4, H5 F
Lee Ct 142-A5 Y
Lee Dr 90-C8 J
Lee Rd 142-K8 H; 142-J8, K8; 143-A7 L; 157-G4 NN; 92-A10; 107-H4, J3; 108-A1 Y
Lee St 159-C7; HE-A6 H
Leeds 88-H4 J
Leeds Terr 158-C3 H
Leeds Way 123-J4 NN
Leeland Ave 174-G1 H
Lees Mill Dr 123-K4 NN
Leeward Way 141-H5 NN
Leftwich Ct 160-D1 H
Legere La 88-H9 J
Lehman Ct 142-K10 H
Leicester Terr 158-B3 H
Leigh Ct 108-F4 Y
Leigh Dr 92-J7 G
Leigh Rd YE-A4 Y
Leigh Terr 125-J8 Y
Leisure Rd 55-H4 J
Leland Pl 124-B10 NN
Lely 88-H6 J
Lemaster Ave 143-K10 H
LeMaster Ave 159-K1 H
Lemburn La 92-F4 G
Lena Cir 158-E6 H
Lenny Dr 155-B8 I
Lenora Dr 141-K7 NN
Lenox Ct 125-C10 Y
Lentz Pl 140-E5 NN
Leo Ct 141-A7 NN
Leon Dr 88-K9; 89-A9 J
Leon La 158-F6 H
Leonard La 141-H8 NN
Leslie Dr 140-K9 NN
Leslie La 141-H3 Y
Lessies Dr 126-K8; 127-A8 Pq
Lester Rd 141-F10; 157-F1 NN
Leta Ct 158-C5 H
Lethbridge La 88-B10 J
Letts Pt Landing 126-H7 Pq
Leusseur Rd 90-H5, K5; 91-A5 Y
Level Green Ct 159-J2 H
Level Way 104-J2 J
Levelfield Pk 125-E5 Y
Levingston Dr 159-K4 H
Levy La 125-J1 Y
Lewallen Dr 124-B7 NN
Lewis Cir 72-G10 J
Lewis Dr 141-A5 NN; 109-K10 Y
Lewis Dr W 158-C7 H
Lewis La 144-C8 H
Lewis Loop 142-K4 H
Lewis Rd E 158-F6 H
Lewis Robert La 89-C3 W
Lexington Cir 141-C4 NN
Lexington Ct 125-D10 Y
Lexington Dr 88-J3 J
Lexington St 159-E1 H
Leyland Ct 141-K5 Y
Libbey St 159-J8 H
Liberty Ch Rd 54-G3 NK
Liberty Cir 141-B3 NN
Liberty Dr 142-A4 Y
Liberty La 158-A3 H
Liberty St 159-D6; HE-B3 H
Lightfoot Rd 72-K6; 73-A6 Y
Lighthouse Dr 144-F8 H
Lighthouse Way 141-E8 NN
Lightning Dr 142-B4 Y
Ligon Pl 124-H6 NN
Lilac Ct 157-H3 NN
Lilburne Way 141-J4 Y
Lillian Ct 143-F10 H
Lily La 88-K7 J
Limetree Ct 124-B9 NN
Linbrook Dr 140-D2 NN
Lincoln St 159-C6; HE-A4, C4 H
Lincolnshire 88-J4 J
Linda Cir 92-H1 G; 159-A5 H
Linda Dr 123-K7; 124-A8 NN
Lindale St 174-J1 H
Lindbergh Rd 142-K6 H
Linden Ave 159-D7; HE-C7 H
Linden Ct 141-K5 Y
Linden La 105-C2 J
Lindrick 88-J4 J
Lindsay Landing 125-K8 Y
Lines Dr 175-J9, K9 NBN
Link Rd 126-B3 Y
Links of Leith 88-F3 J
Linwood Dr 72-G6 J
Lipton Dr 123-K10 NN
Lisa Dr 140-K6 NN
Lisbon Dr 141-G9 NN
Lismore Ct N 141-E2 NN
Lismore Ct S 141-D2 NN
Lite Cir 140-F4 NN
Little Aston 88-G4 J
Little Back River Rd 143-F10; 159-E1 H
Little Back River Rd W 159-E2 H
Little Bay Ave 142-E4 Y
Little Bluff Rd 156-J2 NN
Little Cr Dam Rd 71-D3, D4 J
Little Cr Rd 71-K2 J
Little Dean Cir 124-H6 NN
Little Deer Run 88-C4 J
Little Farms Ave 158-K9; 159-A9 H
Little Florida Rd 127-A10; 142-H1 Pq
Little John Pl 140-E4 NN
Little John Rd 90-D5 Y
Little Neck La 142-G4 H
Little Oak La 143-F10 H
Little Rockwell Way 143-F9 H
Little Round Top 143-H10 H
Live Oak Ct 143-J10 H
Live Oak La 140-J2 NN
Liverpool 88-D3 J
Liza La 89-B1 J
Llewellyn Dr 55-J5 J
Loblolly Ct 109-C8 Y
Loblolly Dr 109-B8 Y
Loch Cir 159-K2 H
Loch Haven Dr 56-E9 J
Loch Ness Dr 124-G9 NN
Lochaven Dr 141-E3, E4 NN
Lochview Dr 141-E4 NN
Locksley Dr 157-J1 H
Lockspur Cres 124-K6 NN
Lockwood Ave 124-H8 NN
Lockwood Dr 159-A7 H
Locust Ave 174-H1 H
Locust La 142-D5 Y
Locust Pl 56-G7 J
Locust Run 127-B10 Pq
Lodge Ct 124-H6 NN
Lodge Rd 143-C1 Pq; 90-F9 Y
Lodi Ct 158-E6 H
Log Cabin Beach Rd 106-J6 J
Logan Ct 159-E1 H
Logan Pl 73-B10 J; 157-E2 NN
Lois La 124-B7 NN
Lolas Dr 140-H6 NN
Lomax St 158-D7 H
Lombard St 158-J10 H
Lombardy Dr 92-H1 G
London Company Way 104-G3 J
Londonderry La 73-E3 Y
Londonshire Terr 158-C2 H
Lone Oak Dr 126-C4 Y
Lonesome Pine Rd 92-G2 G
Long Bridge Rd 159-F1 H
Long Cr Rd 144-D10 H
Long Ct 141-K7 H
Long Green La 160-A4 H
Long Hill 88-J4 J
Longboat 88-D2 J
Longbow Ct 124-G6 NN
Longbridge Way 124-C7 NN
Longfellow Dr 140-F3 NN
Longfellow Rd 107-G6 NN; 107-F3, G5, G6 Y
Longhill Conn Rd 89-B3 J
Longhill Gate 72-G10; 88-G1 J
Longhill Rd 72-E10; 88-J1; 89-B3 J; 89-B3 W
Longhill Sta Rd 72-D8 J
Longhorn Ct 107-C5 J
Longleaf Cir 142-B8 H; 72-J1 J
Longleaf La 124-C3 NN
Longmeadow Dr 141-H7 NN
Longstreet Ct 158-G7 H
Longstreet Rd 140-H6 NN
Longview Dr 92-H2 G
Longwood Cir 125-H6 Y
Longwood Dr 159-E3 H; 157-A3, B2 NN
Lookout Cir 156-H1 NN
Lookout Pass 143-H10 H
Lookout Pt 109-D8 Y
Loon Turn 159-A6 H
Lopez Pl 106-G1 Y
Loquat Pl 158-K1 H
Lorac Ct 89-G4 Y
Lorac Rd 89-G4 Y
Loraine Cir 124-A9 NN
Loraine Dr 124-A9 NN
Lord North Ct 142-A4 Y
Lord Pelham Way 142-F4 Y
Lori Cir 140-G5 NN
Lorigan La 144-D10; 160-D1 H
Loring Ct 158-D7 H
Lorna Doon Dr 125-K7 Y
Lost Cove 141-D7 NN
Lothian 88-J4 J
Lotus St 160-A6 H
Lotz Dr 125-E3, F2 Y
Lou Mac Ct 140-F5 NN
Loudon La 141-G8 NN
Louise Dr 141-G6 NN
Louise La 56-C6, D5 J
Loura Ct 159-A1 H
Lowden Hunt Dr 158-D2 H
Lowell Ave 90-E7 Y
Lowell Pl 140-F3 NN
Lowneyville La 58-J2 G
Lowrey St 159-K7 H
Lowry Pl 124-J6 NN
Loxley La 104-A1 J
Loyal La 140-C3 NN
Luanita La 140-J7 NN
Lucas Cr Rd 124-B8, C9; 140-B1, C2 NN
Lucas Dr 159-E4 H
Lucas Pl 123-E4 F
Lucas Way 142-E10; 158-E1 H
Lucerne Cir 140-K8 NN
Lucinda Ct 158-F2 H
Lucinda Dr 124-C5 NN
Lucy Ct 159-F1 H
Ludlow Dr 109-K9 Y
Lula Carter Rd 108-A6 NN; 108-A6 Y
Lundy La 158-A4 H
Luther Dr 90-E10 Y
Lyas St 159-A8 H
Lyford Key 158-C5 H
Lyliston La 157-G2 NN
Lynch St 91-B4 Y
Lynchburg Dr 140-J8 NN

Lyndon Cir 157-J5 NN
Lynn Cir 126-E9 Pq
Lynn Dr 141-B9 NN
Lynn Lake Ct 39-J6 NK
Lynn Lake La 39-H5 NK
Lynn Lake Rd 39-H5 NK
Lynnette Dr 72-G10 J
Lynnhaven Dr 158-E2 H
Lynns Way 125-F1 Y
Lynnwood Dr 158-K3 H
Lyons Cr Dr 126-H7 Pq
Lyttle Dr 141-A4 NN

M

M Ave 109-G8 Y
M O Herb St 160-B3 H
Mabry Ave 143-D8 L
MacAlva Dr 159-E2 H
MacAuley Bike Tr 106-C2 J
MacAuley Rd 106-B2 J
Mace St 104-F1 J
Machrie 88-E4 J
MacIrvin Dr 140-H10 NN
MacKenzies Run 92-H6 G
MacNeil Dr 140-C1 NN
Macon Ave 157-H6 NN
Macon Cir 73-F5 Y
Macon Rd 158-C5 H
Madeline Pl 141-E8 NN
Maderia Dr 126-D9 Y
Madison Ave 123-G6 F; 157-K8; 158-B10; 174-B1, C2 NN
Madison Chase 142-F7 H
Madison Cir 156-H2 NN
Madison Ct 109-C10; 142-A5 Y
Madison La N 140-J9 NN
Madison La S 140-J10; 156-H2, J1 NN
Madison Rd 90-D9 J
Madrid Dr 143-G10; 159-G1 H
Madrone Pl 158-D2 H
Magazine Rd 89-C5 J
Magistrate La 124-E8 NN
Magna Carta Dr 124-G6 NN
Magnolia Dr 157-H4 NN; 89-B2 W
Magnolia La 143-E1 Pq; 108-G4; YE-A3 Y
Magnolia Pl 159-G5; HE-G2 H
Magnolia St 143-B9 L
Magruder Ave 106-K5; 107-A6 J
Magruder Blvd 142-G9; 158-F1 H
Magruder La 90-B8 J
Magruder Pk 73-G7 Y
Magruder Rd N 157-G4 NN
Magruder Rd S 157-H4 NN
Mahogany La 107-A5 J
Mahogany Run 88-E1 J
Maid Marion Cir 140-G4 NN
Maid Marion Pl 90-C4 Y
Main Rd 91-H7, J8, J9; 92-A10; 108-B1 Y
Main St 157-F6, G5 NN; 108-J3; YE-C2, D2 Y
Mainsail Dr 160-D1 H; 124-D3 NN
Mainship Ct 140-C4 NN
Majestic Ct 140-K9 NN
Majesties Mews 89-F10 W
Mal Mae Ct 90-B9 J
Malbon Ave 141-G5 NN
Malden La 140-D4, E3 NN
Malibu Pl 124-G7 NN
Mall Dr 175-K10 NBN
Mall Pkwy 141-A3 NN
Mallard Cove 88-C10 J
Mallard Cr Run 104-B1 J
Mallard Ct 141-K7 H
Mallard La 157-J8 NN; 109-C9 Y
Mallard Run 159-A6 H
Mallard Run N 72-H9 J
Mallard Run S 72-H10 J
Mallicotte La 156-K2 NN
Mallory St 160-C4, D3 H
Mallory St N 159-H8, K7; 160-B5 H
Malvern Cir 104-K2 J
Malvern Hill Cir 159-H9 H
Mammoth Oak Rd 140-J6 NN
Manack Rd 159-F2 H
Manassas Ct 159-G1 H
Manassas Loop 142-B2 Y
Manchester Dr 158-C6 H
Manchester Run 88-B4 J
Manchester Way 109-B10 Y
Mandy Tr 157-H1 NN
Maney Dr 157-H4, J5 NN
Manhattan Sq 142-G10 H
Manilla La 160-A2 H
Manley Cir 91-J10 Y
Manley Rd 91-K10 Y
Manning La 157-K1; 158-A1 H
Manor Dr E 90-E8 Y
Manor Dr W 90-D8 Y
Manor Hill Ct 157-K2 H
Manor House Ct 109-C8 Y
Manor Pl 140-D6 NN
Manor Rd 124-C8, C9 NN
Mansfield Cir 106-D5 J
Mansford Dr 160-B4 H
Mansion Rd 142-E3 Y
Manteo Ave 159-A9, A10 H
Manton Dr 104-C3 J
Maple Ave 174-F1 H; 174-G1 NN
Maple Ct N 124-B5 NN
Maple Ct S 124-B5 NN
Maple Knoll La 58-H2 G
Maple La 104-A1 J
Maple Rd 108-B5 Y
Maple Rd S 125-A7 Y
Maple St 92-H2 G; 143-C8 L
Maplewood Ct 124-F5 NN
Maplewood Pl 88-H10; 104-J1 J
Maplewood St 159-C5; HE-A3, A5 H
Mara Pk Pl 104-J5 J
Maragret Ct N 125-F4 Y
Maragret Ct S 125-G4 Y
Marauder Dr 142-C7 Y
Marbec La 126-H8 Pq
Marble Run 90-C6 Y
Marcella Rd 158-G3, J2 H
Marclay Rd 105-D1, E1 J
Marcus Dr 141-F5 NN
Marcy Dr 141-H4 Y
Mare Cir 124-E3 NN
Margaret Dr 92-H5 G; 143-F10; 159-G1 H
Margaret Jones Rd 87-J10 J
Margaux Dr 124-D3 NN
Marge Pl 174-H2 H
Marie Cir 158-D4 H
Marie Ct 141-G6 NN; 126-F7 Pq
Marigold La 160-A2 H
Marilea Cir 141-C9 NN
Marina Dr 123-J10 NN
Marina La 140-F7 NN
Marina Pt 73-J2 Y
Marina Rd 159-E8 H; 54-D6 NK
Marine Cir 125-K7 Y
Marine Corps Dr 38-A4 NK
Marine Terminal 174-C4 NN
Mariners Cove Rd 159-G6; HE-F5 H
Mariners Ct 141-A7 NN
Mariners Rd HE-F5 H
Marion 88-F6 J
Marion Rd 160-A4 H
Mark Dr 158-D8 H
Mark Twain Dr 140-F3 NN
Market Dr 158-C10; 174-C1 NN
Market St Ct 104-F2 J
Markham Dr 159-F3 H
Markos Ct 158-D2 H
Marl Borough Ct 158-D6 H
Marl Ravine Rd 109-B8 Y
Marlbank Dr 109-C7, C8 Y
Marlboro Rd 140-D2 NN
Marlbrook Dr 72-D1 J
Marldale Dr 142-J10 H
Marlfield Cir 159-J2 H
Marlin Cir 160-C1 H; 125-J2 Y
Marlin Dr 140-C2 NN
Marly Ct 124-G5 NN
Marmont La 57-B5 J
Maroney La 160-A6 H
Marple La 159-A5 H
Marquette Ct 141-A2 NN
Marrow Dr 141-B7 NN
Marrow St 159-E7; HE-D7 H
Marsh Tacky 90-H8 Y
Marshall Ave 123-G6 F; 159-G7; HE-F8, G7 H; 158-A8, B10; 174-C1, D3 NN
Marshall La 92-K8 G
Marshall Pl 174-E4 NN
Marshall St 159-F5; HE-E1, E2 H
Marshall Way 89-F6; WE-B1 Y
Marshview Dr 124-A6 NN
Marstons La 72-G6 J
Martha Cir 158-A8 NN
Martha Ct 127-C10 Pq
Martha Lee Dr 158-B5, B6 H
Martiau St 108-J3; YE-C2 Y
Martin Ct 105-K4 J
Martin La 159-J5 H; 126-C8 Y
Martin Luther King Jr Blvd 159-H8; HE-H8 H
Martin Rd 141-B9 NN
Martins Ridge 88-G7 J
Martire Cir 141-F9 NN
Marty La 157-J8 NN
Marval Cir 158-A5 H
Marvin Dr 158-K2; 159-A2 H; 124-A6 NN
Marvin St N 142-K6 H
Marvin St S 142-K5 H
Mary Ann Dr 157-J3 H; 110-C8 Y
Mary Bierbauer Way 141-H4 Y
Mary Byrd 88-A7 J
Mary Cir 126-E9 Pq
Mary La 89-B1 J
Mary Peake Blvd 158-E5, E6 H
Mary Robert La 124-C6 NN
Mary St 160-C4 H
Maryland Ave 158-J9 H; 175-J8, J9 NBN
Maryle Ct 140-E3 NN
Maser Ct 158-A6 H
Mashie Ct 125-C10 NN
Mason Ct 109-C10 Y
Mason Dr 124-D4 NN
Mason Row 92-B8 Y
Mason St 159-D5; HE-B1 H
Massacre Hill Rd 107-B6 J
Massasoit La 58-G1 G
Massell Ct 140-H10 NN
Massena Dr 57-B4 J
Massey Ave 141-B8 NN
Massie La 126-C10 Y
Mast Cir 140-C2 NN
Masters La 106-G1 Y
Masters Tr 141-B2 NN
Mastin Ave 125-J1 Y
Mathew Brown 106-A2 J
Mathew Grant 106-A3 J
Mathew Scrivener 106-B2 J
Mathews La 160-A10 FM
Mathews St 108-H2; YE-C1 Y
Matoaka Ave 92-K9 G
Matoaka Ct 89-E6; WE-A2 W
Matoaka La 141-D9 NN
Matoaka Turn 142-B3 Y
Matoka Rd 159-A9 H
Mattaponi Tr 72-J10 J
Matthew Ct 55-D9 J
Matthew Rd 157-C2 NN
Mattie Cir 124-D4 NN
Mattmoore Pl 141-H10 NN
Matton Cir 158-H1 H
Mattox Dr 157-G5 NN
Maume Cir 158-J1 H
Maupin Pl 90-C8 J
Maureen Dr 140-D2 NN
Maury Ave 175-J8 NBN; 157-E1 NN
Maverick Ct 143-J10 H
Maxton La 56-G10; 72-G1 J
Maxwell La 140-G7, J6 NN
Maxwell Pl 106-B2 J
Maxwell St 158-G9 H
May Ct 140-D2 NN
May St 174-J1 H
Mayer Ct 160-D1 H
Mayfair La 88-H2 J
Mayfield Pl 124-H5 NN
Mayfield St 91-D6 Y
Mayland Dr 157-G1 NN
Maymont Dr 141-A6 NN
Maynard Dr 141-G8 NN; 89-K5 W
Maynard St 175-A1 H
Maywood Cir 124-F2 NN
Maywood Ct 158-A2 H
McCall Ct 158-D7 H
McCellan Ave 159-G8; HE-G7, G8 H
McClellan Ct 109-D8 Y
McClurg Dr WE-A2 W
McCrae Dr 124-G4 NN
McCulloch Rd 160-B3 H
McDonald Rd 159-A6 H
McElheney La 159-K2 H
McGuire Dr 142-H7 H
McGuire Pl 141-G8 NN
McKinley Dr 124-A6 NN
McKnew Ct 123-K4 NN
McLain St 123-G6 F
McLaw Dr 123-K4 NN
McLawhorne Dr 157-F3 NN
McLaws Cir 106-D1 J
McMahon St 123-H6 F
McManus Blvd 124-H8, J9 NN
McManus Pl 123-G7 F
McMorrow Dr 124-E6 NN
McMurran Cir 157-B1 NN
McNair Dr 159-K9 FM
McPherson Ct 110-A10 Y
McSweeney Cir 160-A3 H
Meade Dr 140-D5 NN
Meadow Cir 57-B6 J
Meadow Cr Dr 124-F6 NN
Meadow Dr 157-A1 NN
Meadow La 142-D7 H
Meadow Lake Dr 55-G5 J
Meadow Rue Ct 105-G1 J
Meadowbrook 88-H5 J
Meadowbrook Dr 159-A3 H
Meadowcrest Tr 56-F9 J
Meadowfield La 88-H9 J
Meadowfield Rd 108-K9; 109-A10 Y
Meadowlake Rd 141-J2 Y
Meadowlark La 140-G2 NN
Meadowview Dr 125-G7 Y
Meagan Ct 124-A10 NN
Meanwood La 140-E2 NN
Mears Cir 140-J4 NN
Meeting Rd 156-H1 NN
Meghan Kay Cove 140-H9 NN
Meherrin Run 142-C3 Y
Mehrens Ct 160-B3 H
Melba Ct 110-C8 Y
Melbourne Ct 142-F5 Y
Melbourne Pl 159-C8; HE-A8 H
Melena Ct 141-H10 NN
Melinda La 126-B8 Y
Melissa Ct 143-K10 H
Melissa La 73-F2 J
Mellen St 159-J8 H
Mellon St 141-A7 NN
Melody La 109-D10 Y
Melrose Terr 124-G5 NN
Melson La 144-C7 H
Melville Dr 140-F2 NN
Melville Rd 159-A8 H
Melvins End 142-F2 Y
Memorial Dr 141-H3 Y
Memory La 160-C1 H
Menchville Rd 140-F4, F6 NN
Menife Ct 88-G1 J
Mennonite La 140-H3 NN
Menunkatuck La 74-K1 G
Menzels Rd 71-B4 J
Mercantile Dr 158-C10 NN
Mercedes Dr 158-D10 NN
Mercer Rd 92-K10 G
Mercury Blvd 159-F4 H
Mercury Blvd E 159-J6 H
Mercury Blvd W 157-J8; 158-C5, H4; 159-A3 H; 157-J8 NN
Meredith La 125-H7 Y
Meredith Way 140-J5 NN
Meridith St 159-E7; HE-C6 H
Merle Dr 140-F2 NN
Merlyn Walk Pl 140-E1 NN
Merrick Rd 158-E2 H
Merrimac Ave 174-G1 H
Merrimac Ct 158-B7 H; 158-B7 NN
Merrimac Tr 90-C9; 107-A4 J; 90-A5; WE-F1 W; 90-A5, B7, C9; 106-F1 Y
Merritt Dr 126-F9 Pq
Merritt Rd 141-A10 NN
Merry Cir 156-J1 NN
Merry La 156-J2 NN
Merry Oaks Dr 124-A5 NN
Merry Oaks La 55-G7 J
Merry Pt Terr 156-J2 NN
Mesquite Pl 158-J1 H
Messick Rd 143-F1 Pq
Meta Cir 124-C10 NN
Mews La 124-A6 NN
Micale Ave 157-K1 H
Michael Irvin Dr 124-B5 NN
Michael La 92-G3 G
Michael Pl 125-H7 Y
Michael St 142-K10 H
Michaels Woods Dr 157-J1; 158-A1 H
Micheal Ct 107-B7 J
Michelle Dr 143-E10; 159-E1 H; 141-H10; 157-H1 NN
Michelle La 126-F7 Pq
Michelles Ct 72-B2 J
Michigan Dr 159-B6 H
Michlebring La 142-C10 H
Micott Dr 158-F5 H
Mid Ocean 88-C4 J
Middle Ground Blvd 141-C6 NN
Middle Rd 142-D7 H; 110-B7 Y
Middle Woodland Close 103-H1 J
Middleboro Terr 158-K9 H
Middleburg Hunt Rd 158-D3 H
Middlesex Rd 141-B8 NN
Middleton Ct 159-A8 H
Middlewood La 125-H6 Y
Midlands Rd 89-A7 J
Midlothian Sq 159-H2 H
Mikes La 157-J8 NN
Milden Rd 88-K7 J
Mildred Ct 72-E9 J
Mildred Dr 72-E9 J
Mile Course 106-D2 J
Miles Cary Mews 159-F6; HE-E4 H
Miles Cary Rd 140-J10 NN
Miles Mahone 88-F7 J
Milford Ave 159-B7 H
Milford Rd 157-F5 NN
Militia Ct 142-A4 Y
Mill Cr Rd 55-B8 J
Mill Cr Terr 159-J8 H

Mill Crossing 141-K2 Y
Mill La 125-H7, H8 Y
Mill Neck Rd 89-C9, D8 W
Mill Pond Ct 141-C1 NN
Mill Pond Run 55-K1, K2 J
Mill Pt Dr 159-F6; HE-E4 H
Mill Rd 125-G9 Y
Mill Run Ct 107-H9 NN
Mill Stream Way 104-J2 J
Mill View Cir 104-K2 J
Millecent Ct 158-E7 H
Miller Cr La 141-F3 NN
Miller Dr 142-K10 H
Miller Rd 140-B4 NN
Miller Rd E 39-H9 J
Miller Rd W 39-G9 J
Millers Cove Rd 140-B4 NN
Millers Rd 40-E7 NK
Millgate Ct 125-B10 NN
Millside Way 125-J3 Y
Millson Rd 90-K4; 91-A4 Y
Millstone Ct 141-D1 Y
Millwood Dr 124-G9 NN
Miln House Rd 103-C1 J
Milo Dr 41-K4 KQ
Milstead Rd 140-H6 NN
Milton Dr 158-B4 H
Mimado Ct 141-E1 Y
Mimosa Cres 158-H10 H
Mimosa Dr 157-A2 NN; 89-G8 W
Mindspring 88-H9 J
Minetti Ct 142-K9 H
Mingee Dr 158-F8 H
Mingee St 126-K7 Pq
Minor Ct 88-G2 J
Minson Rd 89-B4 J
Minton Dr 141-A9 NN
Minuteman Dr 141-B2 NN
Miranda Ct 159-J4 H
Mirror Lake Dr 56-G10 J
Miss Minnies La 92-B1 G
Mission La 92-G4 G
Missionary Ridge 143-F9; 159-G1 H
Mistletoe Dr 157-A3, B2 NN
Misty Cr Ct 124-D8 NN
Misty Cr La 159-H1 H
Misty Ct 104-H1 J
Misty Dr 126-D2 Y
Misty Pointe La 123-H1 NN
Mitchell Pt Rd 140-J3 NN
Mitchell Rd 159-E3 H
Mitchells Method 142-F2 Y
Mizzen Cir 160-D1 H
Mlle La 104-D3 J
Mobile La 72-H6 J
Mobile Pl 141-A5 NN
Mobjack Loop 142-D1 Y
Mobjack Pl 141-E7 NN
Mockingbird Dr 105-C1 J
Moger Dr 159-J4 H
Mohawk Rd 159-G3 H
Mohea Cir 140-E3 NN
Molesey Hurst 88-C3 J
Moline Dr 141-B8 NN
Mona Dr 124-F6, G7 NN
Monarch Dr 140-F5 NN
Monarch Glade 125-D4 Y
Monarda Ct 123-K9 NN
Moncure Dr 89-B3 J
Monette La 155-D9 I
Monette Pkwy 155-D9 I
Monica Dr 141-B8 NN
Monitor Dr 159-D9 H
Monroe Ave 123-E4, F6 F; 124-G6 NN
Monroe Ct 142-B5 Y
Monroe Dr 159-E3 H
Montague Ct 90-F4 Y
Montclare Pl 140-J9 NN
Monterey Ave 159-B7 H
Monterry Pl 124-G7 NN
Montgomery Dr 92-H5 G
Montgomery La 110-B9 Y
Monticello Ave 87-J10; 88-B8, J7; 89-A6 J; 89-D6; WE-A1 W
Monticello Ct 141-A4 NN
Montrose 88-E4 J
Monty Manor 141-H4 Y
Monument Ct 141-K5 Y
Monument Dr 90-B8 J
Monumental Ave 89-K6; WE-F2 W
Moodys Run 106-C4 J
Mookie Ct 124-E10 NN
Moore Dr 73-B9 J
Moore House Rd 108-J4; 109-A4, B5; YE-D4, G5 Y
Moore La 109-B5 Y
Moore Rd 126-H6 Pq
Moore St 159-B8 H
Moores Cr Dr 126-B9 Y
Moores La 39-B8 NK
Moores La N 140-K10; 157-A1 NN
Moores La S 157-A1 NN
Mooretown Rd 73-C9; 89-D1 J; 89-D1 W; 73-B6, C9; 89-D1 Y
Moray Firth 88-K3 J
Morgan Dr 159-J4 H; 104-J4 J; 157-A1 NN
Morgarts Beach La 155-C5 I
Morgarts Beach Rd 155-A6, B5, C5 I
Moring Ct 124-G8 NN
Morningview Ct 160-D3 H
Morris Cr Cres 86-B2 CC
Morris Cr Landing 86-A2 CC
Morris Dr 157-J7 NN
Morris Farm La 58-J8 G
Morris La 92-F5 G
Morris St 159-K4 H; 175-H9 NBN
Morrison Ave 157-F2, F3 NN
Morton Cir 142-H10; 158-H1 H
Mortor Loop 107-H7 NN
Mosbey Dr 90-C8 J
Mosby Ct 158-D7 H
Moses Harper 103-D1 J
Moses La 90-D9 J
Moss Ave 159-C8 H; 125-H1 Y
Moss Side La 56-K8; 57-A8 J
Mossy Cr Dr 104-B1 J
Motley La 92-K7 G
Motoka Dr 124-F10 NN
Mott La 104-C3 J
Mount Airy Pl 124-G5 NN
Mount Laurel Rd 56-C4 J
Mount Nebo Rd 39-E4 NK
Mount Pleasant Dr 58-A9 J
Mount Pleasant Rd 38-A4 NK
Mount Sterling Cir 159-H9 H
Mount Vernon Ave 89-E5, E6; WE-A1, A2 W
Mount Vernon Ave N 89-E5 W
Mount Vernon Dr 125-J10 Y
Mountain Ash Pl 158-J1 H
Mounts Bay Rd 106-B3 J
Moyer Rd 123-G8, K10 NN
Moyock Run 142-C2 Y
Moysonike Ct 54-E9 NK
Muffin Ct 104-G2 J
Muirfield 88-F3 J
Muirfield Ct 141-D1 Y
Mulberry Ave 174-G3 NN
Mulberry Is Rd 123-D7, E10; 139-F1 F
Mulberry Turn 143-K9 H
Muller La 141-B6 NN
Mullins Ct 141-J10; 157-J1 H
Mullis Ct 92-G1 G
Mumford Cove Rd 92-H5 G
Mumford View Dr 92-J7 G
Municipal Dr 126-K10; 127-A10 Pq
Municipal La 157-G5 NN
Mura Ct 123-K8 NN
Murray Ave 158-K4; 159-A4 H
Murray Ct 143-D5 L
Muse La 108-G5; YE-A5 Y
Museum Dr 157-B1, B2, B4 NN
Museum Pkwy 157-B3 NN
Mushroom Ct 156-J1 NN
Musika Ct 124-E10 NN
Musket Ct 141-C3 NN
Musket Dr 90-B6, C7 Y
Musket Loop 107-H7 NN
Musket Rd 107-J7 NN
Mustang Ct 142-B4 Y
Myers Ct 125-H10 Y
Myers Rd 125-H10 Y
Myles Ct 124-E10 NN
Myra Dr 158-D8 H
Myrtle Ct 124-D3 NN
Myrtle St 159-G5; HE-G1 H
Mytilene Dr 158-C7 H; 158-C7 NN

N

N Ave 109-G8 Y
Nadine Pl 141-H7 NN
Nairn 88-G4 J
Nancy Ct 124-E10 NN
Nancy Dr 159-G1 H
Nansemond Dr 158-B10; 174-B1 NN
Nansemond Turn 142-C3 Y
Nantucket Pl 141-E7 NN
NASA Dr 142-K7 H
Nassau Pl 158-A6 H
Nassau St 89-H7; WE-C3, C4 W
Natalie Cir 124-D4 NN
Natalie Ct 141-K7 H
Nathan St 160-A1 H
Nathaniels Close 103-D1 J
Nathaniels Green 103-D1 J
Nathaniels Run 103-D1 J
National Ave 159-J8 H
National La 90-C6 Y
Naurene Ct 141-H2 Y
Neal Ct 104-C1 J
Neal St 158-K9 H
Nealy Ave 143-B10 L
Neck-O-Land Rd 104-H5 J
Ned Dr 159-A2 H
Neff Dr 143-J9 H
Neighbors Dr 73-A8 J
Nelms Ave 174-E5 NN
Nelms La 104-C1 J
Nelson Ave 89-F6; WE-B2 W
Nelson Cir 90-C8 Y
Nelson Ct 126-G8 Pq; 142-A4 Y
Nelson Dist Rd 108-K9 Y
Nelson Dr 58-K4 G; 157-D3 NN; 90-C7 Y
Nelson House La 159-F1 H
Nelson Pkwy 143-E10; 159-E1 H
Nelson St 108-J3; YE-D3 Y
Neptune Dr 158-H8 H
Neptune Pl 140-C5 NN
Nesbitt Ct 140-K6 NN
Nettles Dr 140-J4; 141-A5, A7 NN
Nettles La 142-G9 H
Nevada Cir 73-E4 Y
Nevalou Ct 73-D9 Y
Neville Cir 159-K3 H
New Castle Dr 88-J10 J
New Day Ct 124-H9 NN
New Dr 157-K6 H
New Garden St 160-B9 FM
New Haven Ct 124-C7 NN
New Hope Rd 89-D5 W
New Kent Ct 141-B2 NN
New Kent Hwy 39-C2 NK
New Market 88-B2 J
New Pt Rd 88-K1 J
New Quarter Dr 89-A7 J
New St 158-G10 H
New York Ave 158-J9 H
Newburn Ave 159-K1 H
Newbury Cir 124-H4 NN
Newby Dr 158-G5 H
Newcastle Ct 158-C6 H; 140-D1 NN
Newcombe Ave 159-E8 H
Newgate Vill Rd 142-B7 H
Newland Ct 104-G2 J
Newman Ct 88-G2 J
Newman Dr 141-H8 NN
Newman Rd 57-F10; 73-F2 J; 73-D4, F2 Y
Newmarket Dr 157-K7 H
Newmarket Dr N 157-K7 NN
Newmoor Dr 92-K10 G
Newport Ave 157-E1 NN; 89-G8, H8; WE-B4, C4 W; 90-C7 Y
Newport News Ave 159-A7; HE-A6 H
News Rd 88-B6, G7, J7 J
Newsome Dr 174-C1 NN; 125-D1 Y
Newsome Pl 158-K7 H
Newton Rd 160-A2, A4 H
Newton St 92-H1 G
Newtown St 159-A8 H
Ney Ct 57-B5 J
Niblik Way 125-C10 NN
Nice Dr 56-D9 J
Nicewood Dr 140-E3, F4 NN
Nicholas Ct 89-B1 J
Nichols Pl 141-A4 NN
Nicholson St 89-J7; WE-D3 W
Nickerson Blvd 160-A2 H
Nicklaus Dr 141-C2 NN
Nicks Landing Cir 126-G6 Pq
Nicol Rd 90-J5 Y
Nicole Pl 157-H1 NN
Nimitz Rd 107-F4 Y
Nina Cir 73-F4 Y
Nina Ct 140-E3 NN
Nina La 72-F3, F4, H3 J
Ninebark Ct 124-C3 NN
Nixon Ave 157-J5 NN
Noble Rd 158-F5 H
Nobles Landing Rd 126-D4 Y
Norfolk St 89-F8; WE-B4 W
Norge Farm La 72-E3 J
Norge La 72-G1 J
Norma Ct 140-G3 NN
Norman Davis Dr 56-C6 J
Norman Dr 126-H8 Pq
Norman St 160-C5 H
Normandy La 140-H6, H7, H8 NN
Normil Landing 140-F6 NN
Normill Landing Dr 140-F6 NN
Norml Rd 140-H6 NN
Norris Ct 158-A5 H
North Armistead Ave 142-G5, H6, J10; 158-J1; 159-A4; HE-A2 H
North Ave 159-K6 H; 157-G7, H6 NN
North Back River Rd HE-A5 H
North Beach Rd 109-F6 Y
North Berwick 88-F2 J
North Boundary St 89-G7; WE-C3 W
North Bowman Terr 141-K1 Y
North Boxwood St HE-G2 H
North Carriage Hill Dr 126-E10 Pq
North Cedar Ct 124-C5 NN
North Chalice Ct 124-E5 NN
North Chase 104-B1 J
North Constitution Dr 125-G4 Y
North Cove Rd 73-F1 J
North Ct 105-C1 J
North Cypress Ct 124-B5 NN
North Dogwood Ct 124-C5 NN
North Dryden St 142-J5 H
North Elm Ct 124-B5 NN
North England St 89-H7; WE-D2 W
North Fallon Ct 159-B7 H
North Fern Cove Ct 142-A2 Y
North Founders Hill 103-B1 J
North Fowlers Close 87-D10 J
North Francis Chapman 104-G1 J
North Fuel Dr 90-K6; 91-A6 Y
North Gawain Way 159-E4 H
North Glenwood Rd 159-E1 H
North Green Dr 141-E4 NN
North Greenfield Ave 158-K2; 159-A2 H
North Henry St 89-G6, H7; WE-C2, C3 W
North Independence Dr 158-J6 H
North Joshua Way 125-G5 Y
North Joyce Lee Cir 158-F6 H
North King St 159-E3, E5 H
North Lake Loop 141-J7 H
North Lakeland Cres 141-K1 Y
North Lawson Rd 127-B8 Pq
North Lismore Ct 141-E2 NN
North Madison La 140-J9 NN
North Magruder Rd 157-G4 NN
North Mallard Run 72-H9 J
North Mallory St 159-H8, K7; 160-B5 H
North Maple Ct 124-B5 NN
North Maragret Ct 125-F4 Y
North Marvin St 142-K6 H
North Moores La 140-K10; 157-A1 NN
North Mount Vernon Ave 89-E5 W
North Newmarket Dr 157-K7 NN
North Oak Ct 124-B5 NN
North Pecan Ct 124-B5 NN
North Pine Ct 124-B5 NN
North Quarter 106-A2 J
North Redwood Ct 124-C5 NN
North Richard Buck 104-H1 J
North Richard Grove 104-H1 J
North Richard Pace 104-H1 J
North Riverside Dr 54-J10; 70-F2 J
North Robert Hunt 104-H1 J
North Rodgers Ave 175-H8 NBN
North Roma Rd 143-C5 L
North School La 125-H6 Y
North Shannon Dr 141-K1 Y
North Spruce Ct 124-B5 NN
North Starfish Ct 159-H4 H
North Stocker Ct 88-G1 J
North Stuart Rd 157-G4 NN
North Trace 72-J9 J
North Trellis Ct 124-B6 NN
North Valasia Rd 126-H10 Pq
North Walker Rd 158-E5 H
North Waterside Dr 38-B6; 54-D1 NK
North Wedgewood Dr 141-K6 NN
North Westover Dr 126-E9 Pq
North Will Scarlet La 90-D4 Y
North Willow Ct 124-B5 NN
North Winona Dr 158-K9 H
Northampton Dr 158-B3 H
Northcutt Dr 160-B4 H
Northhall Way 123-J4 NN
Northpoint Rd 105-G1 J
Northwood Dr 158-G7 H
Norwood Cir 158-K8 H
Nottingham Dr 143-K10; 160-A1 H; 90-E4 Y
Nottingham Tr 140-D5 NN
Nottoway Turn 142-D2 Y
Nurmi La 141-J7 H
Nurney Dr 141-H8 NN
Nuthatch Dr 105-C2 J
Nutmeg Quarter Pl 141-B10 NN

O

O Ave 109-G9 Y
O Canoe Pl 159-A10 H
O Halloran Way 89-A3 J
O Hara La 107-H10 NN
O Keefe La 159-J1 H
Oak Ave 174-E2, F3 NN
Oak Ct 56-D6 J
Oak Ct N 124-B5 NN
Oak Ct S 124-B5 NN
Oak Dr 90-D9 J; 90-D9 Y
Oak Hill Dr 104-J2 J
Oak Lawn Way 55-G5 J
Oak Pt Dr 109-F7 Y
Oak Pt La 109-F7 Y
Oak Rd 105-A2 J; 143-B8 L
Oak Ridge Ct 88-F7 J
Oak Ridge Dr 107-G9 NN
Oak Springs Ct 141-A2 NN
Oak St 141-G2 Y
Oak Tree Farms La 58-J1 G
Oakcrest Dr 159-J4 H
Oakdale Ct 92-H6 G
Oakland Ave 159-H6 H
Oakland Dr 72-D1 J; 157-D5 NN
Oakleaf Ct 124-C3 NN
Oakmont Cir 90-E7 Y
Oakmoore Rd 126-F9 Pq
Oaktree Rd 73-D5, E6 Y
Oakville Rd 143-F10 H
Oakwood Dr 88-F5 J; 142-D5 Y
Oakwood Pl 124-G5 NN
Occoquan Turning 142-D1 Y
Odd Rd 126-J10; 142-J1 Pq
Odessa Dr 143-J10 H
Odum Ct 157-K8 NN
Ogden Cir HE-E7 H
Ogilvy La 156-A10 I
Ohio La 73-E3 Y
Okee Cir 92-K9 G
Old Aberdeen Rd 158-F10 H
Old Ave 158-D9 NN
Old Big Bethel Rd 158-B3 H
Old Briarfield Rd 158-H7 H
Old Bridge Ct 123-J3 NN
Old Bridge Rd 123-J3 NN
Old Buckingham Rd 143-E10 H
Old Buckroe Rd 159-J6; 160-A5, B1 H
Old Carriage Way 88-G7 J
Old Cart Rd 88-F7 J
Old Celey Rd 159-B7 H
Old Ch Rd 72-F2 J
Old Chestnut Ave 158-C9 NN
Old Coach La 124-B8 NN
Old Colonial Dr 72-E9 J
Old Colony La 89-C10 J
Old Courthouse Way 124-D9; 140-E1 NN
Old Dare Rd 125-J4 Y
Old Denbigh Blvd 124-J7 NN
Old Dominion Cir 124-C4 NN
Old Dominion Rd 108-J9; 124-K1 Y
Old Ferry Rd 108-K1 G
Old Field Rd 88-J8 J
Old Field St 160-A6 H
Old Fort Eustis Blvd 124-C3 NN; 108-G10; 124-G1 Y
Old Fox Hill Rd 159-E3 H
Old Glory Ct 90-C4 Y
Old Grist Mill La 140-H4 NN
Old Hampton La 159-E7; HE-C5 H
Old Harpersville Rd 141-H7; 157-E2 NN
Old Hollow Rd 90-F9 Y
Old House Pt Rd 126-D5 Y
Old Jamestown Ct 89-C10 J
Old Lakeside Dr 125-K4, K5 Y
Old Landing Rd 109-C8 Y
Old Lucas Cr Rd 140-C1 NN
Old Mallory Rd 142-D9; 158-D1, E3 H
Old Meribeth Rd 143-F9 H
Old Mill Ct 141-F3 NN
Old Mill La 56-J8 J
Old Mooretown Rd 73-A7 J; 72-K6; 73-A7 Y
Old Neck Rd 70-D5 CC
Old News Rd 88-H7 J
Old Oak Ct 158-C2 H
Old Oak Dr 124-G9 NN
Old Oyster Pt Rd 141-D4 NN
Old Pinetta Rd 58-H4, J4 G
Old Pond Ct 142-C6 H
Old Pond Rd 126-H10 Pq; 125-J5 Y
Old Pt Ave 159-G6, J6; HE-G3 H
Old Railway Rd 125-K3 Y
Old Retriever Tr 70-F4 CC
Old Rte 60 55-E6 J
Old Seaford Rd 110-C10 Y
Old Stable Rd 124-C2 NN
Old Stage Rd 39-F9; 55-G1, J5; 56-B6 J
Old Taylor Rd 73-C8 Y
Old Telegraph Rd 54-A3 NK
Old Town La 159-E6; HE-D5 H
Old Williamsburg Rd 90-K10; 91-A10; 107-B1, E3; 108-A4, D3 Y
Old Wormley Cr Rd 109-D9 Y
Old York Rd 90-H9 Y
Old York-Hampton Hwy 109-A8; 125-D1 Y
Olde Pond La 126-C7 Y
Olde Towne Rd 73-A10; 88-K1; 89-A1 J; 104-C10 SC
Olde Towne Run 123-J4 NN
Oldenburg La 144-C9 H
Oldham Way 123-J4 NN
Oleary Rd 70-A1 CC
Olga Ct 159-F1 H
Olin Dr 140-E2 NN
Olive Dr 141-H7 NN
Olivers Way 73-E2 Y
Olivis Rd 58-K6, K7 G
Olson Ct 158-E2 H
Olympia Pl 158-F1 H
Olympic 88-G5 J
Omera Pl 158-A4 H
Omni Blvd 141-F6 NN
Omni Way 141-G7 NN
Onancock Tr 140-J2 NN
Onancock Turning 142-D1 Y
One Two Three Ct 144-B10 H
Oneda Dr 160-A5 H
Oneonta Dr 124-K7 NN
Onley Dr 125-G10 Y
Onnes Dr 141-C5 NN
Opal Dr 140-G3 NN
Opal Pl 141-A5 NN
Operations Dr 140-K3; 141-A3 NN; 125-F1 Y
Orange Plank Rd 143-F9 H
Orange Rd 90-D10 J
Orangewood Ct 124-G5 NN
Orchard Cir 140-C4 NN
Orchard Rd 159-G7; HE-F8 H
Orchard View 125-E5 Y
Orchards Ave 174-J1 NN
Orcutt Ave 157-K4 H; 158-A8, B9; 174-C1, D2, E3, F4 NN
Oriana Rd 124-E10 NN; 125-B7, F7 Y
Orion Ct 142-B1 Y
Orkney Pl 124-G7 NN
Osage La 143-K9 H
Oscar Loop 141-C8 NN
Oscars Ct 142-G2 Pq
Oser La 159-C8 H
Oslo Ct 72-F2, G2 J
Osprey Ave 174-J1 H
Osprey Dr 70-G2 J
Osprey Pt 109-D9 Y
Osprey Way 124-A5 NN
Otey Dr 70-G2 J
Otley Rd 159-C8 H
Otsego Dr 124-K7 NN
Otter Tr Rd 91-A8 Y
Ottis St 141-E2 NN; 141-E2 Y
Otyakwa Landing 126-K7 Pq
Outpost Rd 54-A5 NK
Overbrooke Pl 159-E2 H
Overlook Cove 140-F7 NN
Overlook Ct 143-H9 H
Overlook Dr 105-H2 J
Overpass Rd 55-H2 J
Overton Dr 159-B2 H
Overton Tr 55-K5 J
Overview Ct 89-G10 W
Owen Davis Blvd 141-K4 Y
Owen St 159-C5 H
Owens Rd 140-A1 NN
Owl Cr Ct 159-E2 H
Oxburgh Pl 124-G6 NN
Oxford Cir 89-B10 J
Oxford Mews 142-H1 Pq
Oxford Rd 89-A10; 105-B1 J; 141-B8 NN
Oxford Terr 159-C9 H
Oxmor Ct 88-H9 J
Oyster Cove Rd 126-B6 Y
Oyster Pt La 141-B4 NN
Oyster Pt Rd 140-J4; 141-A4 NN

P

Pacers Pt 143-K9 H
Pacific Dr 158-F2 H
Pacifica Ct 124-G7 NN
Packetts Ct 106-D2 J
Paddington Ct 124-H4 NN
Paddington Pl 159-A4 H
Paddock Dr 157-B1 NN
Paddock La 143-J9 H; 89-A9 J
Padgett Ct 141-B9 NN
Padgetts Ordinary 106-D4 J
Page Dr 159-J1 H
Page Pl 123-K4 NN; 126-J10 Pq
Page St 89-K7; WE-F2 W
Pageland Dr 126-B10 Y
Pagewood Dr 124-G9 NN
Paige Rd 38-B1 NK
Palace Ct 158-C5 H; 124-F4 NN
Palace La 89-F5 Y
Palace St 89-H7; WE-D3 W
Palamino Dr 142-D7 H
Palen Ave 157-F5 NN
Palisade Pt 157-J3 H
Palladium Pl 141-G8 NN
Palmer Ct 125-F1 Y
Palmer La 141-B3 NN
Palmer St 159-C6 H
Palmersto Dr 159-C8; HE-A8 H
Palmerton Dr 124-J7 NN
Palomino Dr 140-F1 NN
Pam La 140-D2 NN
Pamela Dr 158-F6 H; 141-G10 NN
Pamela Pl 125-J2 Y
Pamlico Run 142-C2 Y
Pamunkey Turn 142-B2 Y
Pan Ct 141-J6 NN
Pansy St 160-A6 H
Panther Paw Path 90-E8 Y
Papadam Ct 142-E2 Y
Paquette Ct 157-K1 H
Par Dr 88-E3 J
Paradise Pt Rd 126-B2 Y
Paradise Way 108-A7 NN
Parchment Blvd 90-F10; 106-F1 Y
Parish Ave 174-F2 NN
Park Ave 174-H2 NN
Park Cir 90-E10 Y
Park Hill Cir 140-C4 NN
Park La 110-A10 Y
Park Pl 159-E7; HE-C8 H; 157-F6 NN
Park St 127-B10 Pq
Parkdale Ave 159-B6 H
Parke Ct 88-F1 J; 89-E10 W
Parker Ave 140-J7 NN
Parker La 110-A8 Y
Parkinson Rd 158-F8 H
Parkside Ave 159-D4 H
Parkside Dr 124-D3 NN
Parkside La 87-F10 J
Parkview Dr 157-K9 NN
Parkview Dr S 157-K9 NN
Parkview Pl 160-D3 H
Parkway Dr 156-K3; 157-A2 NN; 89-K6; 90-A7; WE-E1, F3 W
Parkway Rd 159-D3 H
Parkwood Dr 142-D5 Y
Parliament La 124-E7 NN
Parma Ct 124-B10 NN
Parrish Ave 174-E1 H
Parsonage La 159-F1 H
Partridge Ct 92-H1 G
Partridge Landing Rd 41-J4 KQ
Pasadena Ct 142-J10 H
Pasbehegh Dr 104-B1 J
Paschal Pl 156-A10 I
Paspeheghe Run 142-C2 Y
Pasture Cir 57-B7 J
Pasture La 159-C2 H; 142-D2 Y
Pasture Rd 126-H7 Pq
Pat Lloyd Ct 90-J7 Y
Patch Rd 123-E5 F; 160-A9 FM
Patel Way 157-G2 NN
Pates Cr 105-D4 J
Patricia Dr 123-K4 NN
Patrician Dr 158-C2 H
Patrick Ct 159-C5 H
Patrick Henry Dr 124-H8, J10 NN; 89-C2 W
Patrick La 123-H9 NN; 126-F8 Pq
Patrick St 159-C5 H
Patricks Cr Rd 125-K5 Y
Patriot Cir 141-B3 NN
Patriot Cres 157-K4 H
Patriot Way 142-A4 Y
Patriots Colony Dr 88-A10 J
Pats La 157-K9 NN
Patterson Ave 159-C5, C6 H
Patton Ave 123-H5, J6 F
Patton Dr 140-J7 NN
Patton Rd 160-B9 FM
Patuxent Turn 142-D2 Y
Paul Dr 141-F9 NN
Paul Jack Dr 158-G1 H
Paul St 158-A8 NN
Paula Ct 141-H4 Y
Paula Dr 124-A7 NN
Paula Maria Dr 156-H1 NN
Paulette Dr 124-B9 NN
Pauline Cir 140-F4 NN
Pauline Dr 159-J5 H
Pauls Pk Cir 124-C8 NN
Paulson La 160-C1 H
Pavilion Pl 144-D10 H; 141-D8 NN
Paynes Rd 90-F9 Y
Pea Ct 73-B2 Y
Peabody Dr 158-G2 H
Peacepipe Pl 90-F10 Y
Peach St 72-G1, H1 J
Peach Tree Cres 140-D5 NN
Peachtree 88-F5 J
Peachtree La 159-F2 H; 141-J3 Y
Peachwood Ct 124-G5 NN
Peale Ct 88-G1 J
Pear Ave 174-G1 H; 174-G1 NN
Pear Ridge Cir 124-D10; 140-D1 NN
Pear Tree Ct 124-C9 NN
Pearcewood La 160-C1 H
Pearl Cir 140-F2 NN
Pearl Pl 123-F6 F
Pebble Beach 159-B2 H; 88-H5 J
Pebble Beach La 106-G1 Y
Pecan Ct N 124-B5 NN
Pecan Ct S 124-B5 NN
Pecan Rd 158-B5 H
Peebles Dr 124-J9 NN
Peek St 159-E7; HE-C7 H
Peirsey Pl 123-J9 NN

71

Royston Dr 159-F4 H
Rozzelle Rd 159-J3 H
Ruby Ct 159-A1 H
Ruby La 124-C8 NN
Ruckman Rd 160-A10 FM
Rudd La 159-E7; HE-D6 H
Rudisill Rd 143-G10; 159-F1 H
Rue De Rayborn 123-H2 NN
Rue Degrasse 126-H8 Pq
Rugby Rd 141-D8 NN
Rumson Ave 141-G7 NN
Run Way 125-B7 Y
Runaway La 108-K8 Y
Runey Way 141-F1 Y
Running Cedar Ct 108-F4 Y
Running Cedar Way 88-J10 J
Running Man Tr 142-B3, C1 Y
Running Path 160-B9 FM
Rural Retreat Rd 125-H8 Y
Rushlake Ct 140-H4 NN
Russell Ct E 158-C10 NN
Russell Ct W 158-B10 NN
Russell La 142-A4 Y
Russell Rd E 158-F6 H
Rust St 160-B5 H
Rustling Oak Ridge 125-F5 Y
Ruston Dr 124-F10 NN
Rusty Ct 90-D7 Y
Ruth Ct 157-K2 H; 124-A10 NN
Ruth Dr 126-G7 Pq
Ruth La 72-G7 J
Rutherford Dr 57-E9 J
Rutherford Rd 141-B5 NN
Rutherford St 159-A7 H
Rutledge Rd 157-E2 NN
Ryan Ave 143-D5 L
Ryans Run 124-F3 NN
Ryans Way 126-A9 Y
Rye 88-D3 J
Ryland Rd 158-J8 H

S

Sabre Ct 142-B4 Y
Sabre Dr 104-E1 J
Sacramento Dr 142-J10 H
Sacramento Rd 108-E1 Y
Saddle Dr 140-D5 NN
Saddle La 142-D7 H
Saddler Dr 124-B8 NN
Saddletown Rd 57-D9 J
Sagamore 88-H5 J
Sage Ct 159-F7; HE-F6 H
Saint Albans Dr 159-F3 H
Saint Andrews Dr 88-D1, D3 J
Saint Andrews La 124-G8 NN
Saint Annes 88-K3 J
Saint Ashley Pl 160-B1 H
Saint Clare Dr 124-H8 NN
Saint Croix Dr 140-C3 NN
Saint Egnatios Dr 141-J6 NN
Saint Francis Dr 140-C4 NN
Saint George Dr 142-F4 Y
Saint George St 124-J7 NN
Saint George Way 159-A4 H
Saint Georges 88-C3 J
Saint Georges Blvd 104-F1 J
Saint James Pl 124-J7 NN
Saint Johns Ct 90-C5 Y
Saint Johns Dr 158-G5 H
Saint Jude Cir 107-C5 J
Saint Lo Ct 140-H8 NN
Saint Michaels Way 141-C8 NN
Saint Paul Ct 159-A1 H
Saint Simone Ct 89-F10 W
Saint Stephens Dr 124-F10 NN
Saint Thomas Dr 141-B6 NN
Saint Tropez Dr 140-C3 NN
Saint Yves Cir 141-H7 NN
Salem Ct 125-J3 Y
Salem St 159-D3 H
Salina St 159-D3 H
Salisbury Way 159-E5 H; 142-F5 Y
Sally Ann Pl 124-K6 NN
Salt Marsh Quay 158-C2 H
Salt Pond Pl 141-C1 NN
Salt Pond Rd 160-C1 H
Salt Water La 144-B8 H
Salters Cr Rd 158-J8, J9 H
Salters St 174-E2 NN
Samantha Ct 159-J4 H
Samoset 88-J5 J
Sampson Ave 174-G1 H
Samuel Mathews 88-C7 J
Samuel Sharpe 106-A4 J
San Angle Ct 107-B6 J
San Jose Dr 141-F8 NN
Sanctuary Cove 126-B10 Y
Sanctuary Dr 103-G1 J
Sand Dr 56-D6 J
Sand Dr W 56-D6 J
Sand Hill La 92-H6 G
Sand Hill Rd 56-D7 J
Sand Pebble Cir 141-A6 NN
Sanda Ave 91-B5, D5 Y
Sandalwood La 142-D2 Y
Sandbox La 110-C5 Y
Sanderling Walk 125-E4 Y
Sandhurst Cir 141-H7 NN
Sandpiper Cove 126-D4 Y
Sandpiper Ct 143-J10 H
Sandpiper St 140-C2 NN
Sandra Ct 160-A1 H
Sandra Dr 124-A8 NN
Sandstad Ct 72-F4 J
Sandy Bay Dr 126-G5 Pq
Sandy Bay Rd 104-H2, H3 J
Sandy Hill Ct 92-F3 G
Sandy Lake Dr 142-A7 H
Sandy Pt Rd E 126-J7 Pq
Sandy Pt Rd W 126-J7 Pq
Sandy Pt Ridge 88-F7 J
Sanford Dr 158-F8 H; 157-F2 NN
Sangaree Twist 125-H3 Y
Sanlin Dr 140-G4 NN
Sanlun Lakes Dr 142-A7 H
Santa Barbara Dr 142-J9; 143-A10 H
Santa Clara Dr 142-K10 H
Sarah Ct 141-A7 NN
Sarah Spence 105-E4 J
Sarazen Ct 141-C2 NN
Sarfan Dr 160-D2 H
Sargeant Rd 160-A5 H
Sasha Ct 89-B1 J
Sassafras Ct 105-C2 J
Sassafras La 58-G5 G
Sassafras Landing Rd 58-G4 G
Sassafras Rd 58-H5 G
Satinwood La 140-F1 NN
Satterfield Lewis Dr 141-A5 NN
Sauder Way 123-H2 NN
Saunders Dr 126-J8 Pq
Saunders Rd 141-J5 NN
Saunton Links 88-C3 J
Savage Dr 140-G2 NN
Savannah Ct 140-J5 NN
Savannah Sq E 89-F4 Y
Savannah Sq W 89-E4 Y
Saville Row 158-J4 H
Saw Mill Rd 87-J8 J
Sawgrass Turn 125-E5 Y
Saxon La 141-G8 NN
Saxon Rd 90-C5 Y
Saxony Pl 159-E4 H
Saxton Dr 159-H5 H
Saybrooke Ct 140-K7 NN
Scarborough La 142-B10 H
Scarborough Mews 89-A2 J
Scenic Ct 107-G9 NN
Schenck Dr 73-F7 J
Scher Ct 158-C7 H
Schley Ave 174-G1 H
Schmidt Dr 89-B4 J
School La 56-A10 J; 125-H7 Y
School La N 125-H6 Y
Schooner Dr 143-J9 H; 140-C2 NN
Schultz Pl 123-F6 F
Scollin Cir 159-K3 H
Scones Dam Rd 157-J8 NN
Scones Dr 158-E6 H
Scot Ctr Entrance 141-B4 NN
Scotch Pine Ct 124-D3 NN
Scotch Tom Way 125-H3 Y
Scotland Rd 159-J6 H
Scotland St 89-G7, H7; WE-B3, C3, D3 W
Scott Dr 158-D9 H; 90-C7 Y
Scott Rd 140-J10 NN
Scottland Terr 156-K2 NN
Scotts Pond Dr 73-A10 J
Scotts Pt 159-K6 H
Scufflefield Rd 140-H5, J5 NN
Scuttle La 141-K3 Y
Sea Breeze Ct 159-J3 H
Sea Cove Ct 143-H9 H
Sea Pine La 124-B3 NN
Seabee Ave 175-K10 NBN
Seabee Pt 143-G9 H
Seaboard Ave 160-C4 H
Seabreeze Farm La 144-B9 H
Seafarer Cove 158-J6 H
Seaford Rd 109-K10; 110-A10, D10; 125-G1 Y
Seaford St 159-J6 H
Seagrams Ct 141-F1 Y
Seagull Ct 159-F4 H; 124-A5 NN
Searcy Dr 142-H7 H
Seashell Dr 159-H3 H
Seasons Ct 72-J9 J
Seasons Tr 140-G4 NN
Seasons Trace 72-H10, J9, J10 J
Seaview Dr 160-B5, C5 H
Seaward Dr 159-J3 H
Seawells Pt La 92-E1 G
Secluded Pl 123-K6 NN
Secota Dr 158-K10 H
Secretariat La 142-G6 H
Security Dr 107-B9 J
Sedgefield Dr 157-G3 NN
Seekright Dr 141-K4 Y
Segar St 159-H8 H
Selby La 90-D9 J
Selden Rd 156-K2; 157-A2 NN
Seldendale Dr 159-C3 H
Selkirk Dr 140-C1 NN
Seminary Ridge 159-H1 H
Seminole 88-G5 J
Seminole Rd 159-B8 H
Seminole St 92-H4 G
Semple Farm Rd 142-D6 H
Semple Rd 90-C8 Y
Semple Rd W 90-C8 Y
Semple St 159-J7 H
Sentry Cir 125-D2 Y
Serene View Rd 92-E3 G
Service Dr 125-F1 Y
Sesso Dr 160-B2 H
Seth La 142-E1 Y
Seton Hill 88-K1 J
Settlers La 72-G6 J
Settlers Landing Rd 159-C7; HE-A5, F5 H
Settlers Landing Rd E HE-F5 H
Settlers Landing Rd W HE-A5 H
Settlers Rd 156-H1 NN
Seven Hollys Dr 126-B2 Y
Seven Oaks 88-D2 J
Severn Rd 140-J2 NN
Severn St 159-E3 H
Seward Dr 159-K3 H
Sewell Ave 159-G8, J7; HE-G8 H
Shackleford Rd 126-C8 Y
Shackleton La 88-K8 J
Shade Tree Dr 157-E3 NN
Shadwell Ct 141-F8 NN
Shady Bluff Pt 73-H3 Y
Shady Cir 158-D2 H
Shady Grove Pl 158-D2 H
Shady La 88-K8 J
Shady Pond Rd 54-A4 NK
Shady Terr 124-C7 NN
Shadywood Dr 140-D4 NN
Shallow Lagoon 141-H2 Y
Shamrock Ave 141-G2 Y
Shamrock Ct 140-H7 NN
Shanna Ct 125-G3 Y
Shannon Dr 123-K10; 124-A10 NN
Shannon Dr N 141-K1 Y
Shannon Dr S 141-K1 Y
Shapiro Ct 140-H10 NN
Sharon Bass Dr 144-C9 H
Sharon Ct 158-G5 H
Sharon Dr 140-G2 NN; 126-A1 Y
Sharpe Cir 91-H10; 107-H1 Y
Sharpe Rd 107-J1 Y
Sharpley Ave 157-K5 H
Sharps La 90-E10 Y
Sharps Rd 88-K8 J
Shasta Dr 123-K9 NN
Shatt Dr 174-H2 H
Shaughanassee Ct 158-E6 H
Shawen Dr 159-E2 H
Shawn Cir 140-G1 NN
Shea La 90-D8 Y
Sheehan Rd 90-H6 Y
Sheffield La 142-D4 Y
Sheffield Rd 88-H1 J
Sheffield St 158-B2, C2 H
Sheffield Way 140-D1 NN
Sheila Cir 124-A9 NN
Sheila Dr 160-B2 H
Shelby Ave 143-H8 H
Shelby Dr 123-K9 NN
Shell Rd 158-F10, J9; 159-A8, C7; 174-E1 H
Shellbank Dr 103-K2 J
Shelley Ave 157-H6 NN
Shelley Ct 159-J2 H
Shelter Cir 124-D4 NN
Shelton Rd 159-K5 H
Shembri Dr 141-E2 Y
Shenandoah Ct 72-E9 J
Shenandoah Dr 140-G4 NN
Shenandoah Rd 159-B8, B9 H
Shenk Rd 140-D6 NN
Shepard Blvd 142-H8 H
Sheppard Dr 90-B7 Y
Sheppard Pl 123-G5 F
Sheralyn Pl 158-E7 H
Sheraton Terr 158-B4 H
Sherbrooke Dr 140-D5 NN
Sheriffs Pl 90-E5 Y
Sherman Ct 142-A5 Y
Sherry Cir 124-B8 NN
Sherry Dell Dr 157-K3 H
Sherwood Ave 159-K6 H
Sherwood Dr 92-K9 G; 90-F5 Y
Sherwood Forest Rd 58-A9 J
Sherwood Pl 140-F1 NN
Sherwood Rd 91-B3 Y
Shetland Ct 142-D7 H
Shield La 109-D9 Y
Shields Poynt 88-F7 J
Shields Rd 124-E4, G6 NN
Shields St 159-D7; HE-B8 H
Shifting Log Dr 143-G10 H
Shiloh Pk 159-H1 H
Shiloh Pl 174-E4 NN
Ship Pt Rd 126-C3 Y
Ships Landing 141-E7 NN
Shipyard Dr 157-J9 NN; 90-J8 Y
Shire Chase 140-F1 NN
Shirley Ave 89-F6; WE-A1 W
Shirley Dr 159-A2 H; 58-A9 J; 126-B9 Y
Shirley Rd 157-E5 NN; 110-C8 Y
Shnnecock 88-F4 J
Shoal Cr 88-D9 J
Shoe La 141-B10; 156-K2; 157-A1 NN
Shoemaker Cir 140-E6 NN
Shoffel Rd 107-F4, F5 Y
Shooting Star Dr 142-B4 Y
Shore Dr 105-C1 J; 174-F5 NN
Shore Pk Dr 140-B2 NN
Shore Rd 159-F7; HE-E7, G8 H
Shorecrest La 144-A8 H
Shoreham La 90-E6 Y
Shoreline Dr 159-C2 H; 126-E8 Pq
Shoreline Pointe 141-E2 NN
Short Hole 88-K3 J
Short St 174-F2 NN
Shorts La 159-J8 H
Shoveler Ct 159-A6 H
Showalter Rd 125-H8, J7 Y
Shreck Dr 158-C6 H
Shrewsbury Sq 73-A8 J; 125-G6 Y
Shrike Ct 142-C4 Y
Shupper Dr 107-G6 NN
Shuttle Ct 142-G4 H
Sibby Ct 124-F10 NN
Sidewinder Ct 107-B6 J
Sidney Pl 157-G5 NN
Siege La 108-J9 Y
Siemens Way 124-J9 NN
Sierra Dr 141-F8 NN
Signature Way 158-H2 H
Signi Hi Ct 141-F9 NN
Signpost Rd 89-A10 J
Sijan Rd 143-F7 L
Silarmo La 72-F2 J
Silk Tree Pl 158-C1 H
Silky Way 58-H8 G
Silver Fox Trace 125-F10; 141-F1 Y
Silver Isles Blvd 144-D10; 160-C1 H
Silversmith Cir 124-H6 NN
Silverwood Dr 123-J10 NN
Simmons Dr 109-K9 Y
Simon Cir 140-K7 NN
Simone Ct 141-H8 NN
Sinclair La 126-D10; 142-D1 Y
Sinclair Rd 159-D1 H
Singleton Dr 142-C10 H
Sinton Rd 141-F10 NN
Sir Albert Ct 124-F4 NN
Sir Arthur Ct 140-E4 NN
Sir Francis Wyatt Pl 156-K2 NN
Sir John Way 110-A9 Y
Sir Lionel Ct 124-F4 NN
Sir Ralph La 143-A1 Pq
Sir Thomas Lunsford Dr 89-E10 W
Sir Thomas Way 104-F3 J
Sir Walter La 159-A9 H
Sitka Ct 158-B5 H
Six Mount Zion Rd 56-B6, C3 J
Sixpence Ct 90-E5 Y
Skalak Dr 89-B1 J
Skiffes Blvd 107-C7 J
Skiffes Cr Cir 107-C7 J
Skiffs Cr Landing Rd 107-H10 NN
Skillman Dr 55-E4 J
Skimino Landing Ests 73-J2 Y
Skimino Rd 73-H4; 74-B4 Y
Skipjack Rd 140-B3 NN
Skipper Ct 143-H8 H; 140-C2 NN
Skipper La 142-B3 Y
Skyland Dr 160-A4 H
Skyrider Ct 142-C7 Y
Skytrain Dr 142-C4 Y
Slater Ave 160-B4 H
Sleepy Hollow La 125-F2 Y
Sloane Pl 140-K8 NN
Sloane Sq 88-G1 J
Sloop Ct 158-F4 H
Sluice Pond Way 140-H3 NN
Smiley Rd 159-K4 H
Smiroldo Rd 91-C5 Y
Smith Cir 123-G6 F
Smith La 157-E2 NN
Smith St 159-A8 H; 142-J1 Pq; 108-J3; YE-D3 Y
Smithman Cir 159-J2 H
Smokehouse La 104-K4 J
Smokey Tr 140-A1 NN
Smoots La 110-D10; 126-D1 Y
Smucker Rd 140-G2 NN
Smyth Rd 143-C5 L
Snead Dr 141-D2 NN
Snidow Blvd 124-B7 NN
Snodgrass La 157-J4 H
Snow St 159-K6 H
Snug Harbor Dr 159-A10 H
Snug Harbor La 140-G7 NN
Soho St 159-E8 H
Sojourner Ct 140-H3 NN
Soldiers Home Rd 159-G7; HE-F7 H
Solebay St 125-G6 Y
Solo Rd 158-D10 NN
Somerset Cir 125-G2 Y
Somerset Ct 88-J2 J
Somerset La 159-D8 H
Somerset Pl 141-C2 NN
Somervell St 123-G6 F
Somerville Dr 159-K4 H
Sommerville Way 109-K9 Y
Songbird Tr 125-E5 Y
Sonny Hutchins Dr 142-H6 H
Sonoma Woods Dr 124-G8 NN
Sonora Dr 143-K10 H
Sonshine Way 108-A5 Y
Sourwood Dr 158-B6 H
South Armistead Ave 159-D7; HE-C3, C4 H
South Ave 157-G7, H7 NN
South Back River Rd HE-A6 H
South Boundary St 89-G7, G8; WE-C4 W
South Bowman Terr 141-K1 Y
South Boxwood St HE-G3 H
South Cedar Ct 124-C5 NN
South Chalke Ct 124-E5 NN
South Chase 104-B1 J
South Cove Ct 73-F1 J
South Ct 105-C1 J
South Cypress Ct 124-B5 NN
South Dogwood Ct 124-C5 NN
South Dryden St 142-H5 H
South Elm Ct 124-B5 NN
South Emmaus Rd 126-G8 Pq
South England Cir 105-J1 J
South England St 105-J1 J; 89-J7; WE-D4 W
South Fallon Ct 159-B7 H
South Fern Cove Ct 142-A2 Y
South Founders Hill 103-B1 J
South Freeman Rd 103-G1 J
South Fuel Dr 90-K6; 91-A6 Y
South Gawain Way 159-E4 H
South Greenfield Ave 158-K2; 159-A2 H
South Henry St 89-G9; WE-C4 W
South Independence Dr 158-J6 H
South Joshua Way 125-H5 Y
South Joyce Lee Cir 158-F6 H
South King St 159-E7; HE-D5 H
South Lake Cir 141-K7 H
South Lawson Rd 127-D10 Pq
South Lismore Ct 141-D2 NN
South Madison La 140-J10; 156-H2, J1 NN
South Magruder Rd 157-H4 NN
South Mallard Run 72-H10 J
South Maple Ct 124-B5 NN
South Maple Rd 125-A7 Y
South Maragret Ct 125-G4 Y
South Marvin St 142-K5 H
South Moores La 157-A1 NN
South Oak Ct 124-B5 NN
South Parkview Dr 157-K9 NN
South Pecan Ct 124-B5 NN
South Pine Ct 124-B5 NN
South Redwood Ct 124-C5 NN
South Richard Buck 104-H2 J
South Richard Grove 104-H2 J
South Richard Pace 104-H2 J
South Riverside Dr 70-G2 J
South Robert Hunt 104-H2 J
South Rogers Ave 175-H9 NBN
South Roma Rd 143-C6 L
South Shannon Dr 141-K1 Y
South Spruce Ct 124-B5 NN
South St 159-C8; HE-A8 H
South Starfish Ct 159-H4 H
South Stocker Ct 88-G2 J
South Stuart Rd 157-H4 NN
South Terrace Dr 126-J10 Pq
South Trellis Ct 124-B6 NN
South Valasia Rd 126-G10 Pq
South Warner St 142-J6 H
South Waterside Dr 54-C6, D3 NK
South William 106-A3 J
South Willow Ct 124-B5 NN
South Wright St 142-J6 H
Southall Lainding 160-D1 H
Southall Rd 106-B1, D2 J
Southampton Ave E 159-D8 H
Southampton Ave W 159-C8 H
Southeast Trace 72-K9 J
Southern Hills 88-F6 J
Southerland Dr 159-G4 H
Southgate Rd 141-C4 NN
Southhall Way 123-J4 NN
Southlake Pl 140-G4 NN
Southpoint Dr 105-G3 J
Southport Tr 88-B10; 104-B1 J
Southwark Rd 103-D10 SC
Southwind Dr 159-J2 H
Southwold Ct 104-C1 J
Souverain Landing 125-H6 Y
Spanish Tr 159-E5; HE-D1 H
Sparks Ct 89-E9 W
Sparrer Rd 110-C10; 126-C1 Y
Sparrow Ct 107-B7 J
Spatz Dr 143-B9 L
Spaulding Dr 140-C1 NN
Speakers Ct 89-G5 Y
Species La 58-H3 G
Spencer Cir 174-B1 NN
Spencer La 125-E3 Y
Spencer Rd 109-A1 G
Sperry Ct 159-E1 H
Spinnaker Rd 140-C2 NN
Spinnaker Way 58-H9 G
Spitfire Dr 142-C7 Y
Spivey La 126-F1 Y
Split Rail Cir 124-G9 NN
Spoon Ct 125-E10; 141-E1 Y
Spotswood Cay 105-E1 J
Spotswood St 89-H7; WE-D3 W
Spottswood La 156-K2 NN
Spottswood Pl 159-B10 H
Spratley Cir 140-G1 NN
Spring Branch 104-J1 J
Spring E 72-J10 J
Spring House Way 124-G9 NN
Spring Meadow Ct 157-K3 H
Spring Rd 105-A1 J; 141-A7 NN; 107-J2; 108-A2, D5 Y
Spring Trace 72-J9 J
Spring Trace La 141-K6 NN
Spring W 72-H10 J
Springdale Dr 123-J10 NN
Springdale Way 158-D3 H
Springfield Ave 159-C6; HE-A3, B3 H
Springfield Dr 90-D7 Y
Springfield Rd 90-J7 Y
Springhill Dr 88-E6 J
Springmont Ct 124-E7 NN
Springwell Pl 124-J5 NN
Springwood Pl 158-D1 H
Spruce Ct N 124-B5 NN
Spruce Ct S 124-B5 NN
Spruce Dr 142-E5 Y
Spruce Rd 157-G3 NN
Spruce St 158-G9 H; 143-B9 L
Spur Ct 141-B8 NN; 90-F4 Y
Spur Dr 140-K4; 141-A4 NN
Spyglass 88-H5 J
Squires Pl 140-J5 NN
Squires Way 39-G10 J
Stacey Cir 159-K2 H
Stacis La 124-F4 NN
Stadium Dr 89-F7; WE-B3 W
Stafford Dr 124-B6 NN
Staffordshire La 89-B8 J
Stag Terr 141-F5 NN
Stage Rd 38-A2, B3, G3; 39-A4, C8 NK; 140-H10; 156-H1 NN
Stagecoach Watch 125-F4 Y
Staghorn Ct 158-A2 H
Stallion Ct 142-E4 Y
Stanford Ct 142-K9 H
Stanford Pl 104-A1 J
Stanley Dr 88-K8; 89-A9 J; 123-K6 NN
Stanton Ct 159-E3 H
Stanton Rd 140-K10 NN
Staples 106-D2 J
Staples Rd 106-D2 J
Star St 91-C5 Y
Starboard Cir 140-B2 NN
Stardust Cir 124-E9 NN
Starfighter Ct 142-C4 Y
Starfish Ct N 159-H4 H
Starfish Ct S 159-H4 H
Starling Dr 105-C2 J
Starvation Rd 57-K1 G
State Pk Dr 144-F8 H
State St HE-A8 H
States Dr 157-K8 NN
Staton Dr 158-A5 H
Stavenger Ct 72-G3 J
Steakhouse La 142-D4 Y
Steckys Twist 125-F4 Y
Stedlyn Cir 144-D10 H
Steelers Cir 157-K1 H
Steeplechase Dr 88-H7 J
Steeplechase Loop 158-D3 H
Steeplechase Way E 88-H8 J
Steeplechase Way W 88-H8 J
Steffi Pl 140-H9 NN
Stella Cir 142-B1 Y
Stella June La 142-F6 H
Stephen Conway Ct 157-J2 H
Sterling Ct 159-F3 H; 142-A3 Y
Sterling St 157-J6 NN
Sternberg Ave 123-H4 F
Steven Ct 158-F6 H
Stevens Ct 142-A2 Y
Stevens Rd 123-J9 NN
Stewart Dr 90-D8, D9 Y
Stewart Rd 40-B3 NK
Stewart St 159-D6, D7; HE-B5, B6 H
Stewarts Rd 54-H2, J2; 55-A2 J
Stickman Dr 141-E2 NN
Still Harbor Ct 143-H9 H
Stillwater Ct 141-C1 NN
Stillwater La 110-A7 Y
Stillwell Dr 160-A9 FM
Stillwell St 123-J4 F
Stirrup Ct 160-C2 H
Stocker Ct N 88-G1 J
Stocker Ct S 88-G2 J
Stockton St 160-A1 H
Stoke Poges 88-C3 J
Stokes Dr 92-J6 G
Stone Bridge 88-E3 J
Stone Cove 141-F3 NN
Stone Lake Ct 141-K2 Y
Stonegate Ct 141-F3 NN
Stonehouse Rd 57-B6 J
Stonehurst Rd 159-G4 H
Stonewall Pl 140-J6 NN
Stonewall Terr 158-C3 H
Stoney Cr Dr 104-C1 J
Stoney Cr W 104-B1 J
Stoney Pt Rd 108-E1 Y
Stoneybrook La 124-A7 NN
Stony Dr 124-B10 NN
Stony Ridge Ct 124-A8 NN
Stowe 88-H4 J
Stoweflake 88-D1 J
Stratford Dr 89-F5; WE-B1 Y
Stratford Rd 159-D9 H; 89-C1 J; 157-E5 NN
Stratton 88-H5 J
Stratton Major Rd 41-H2 KQ
Strawberry Banks Blvd 159-H8 H
Strawberry La 140-A1 NN
Strawberry Plains Rd 89-B8 J; 89-B8 W
Strock Rd 90-H6 Y
Strong Cir 123-H7 F
Strother Dr 158-A5 H
Stroup Ct 160-B3 H
Stuart Cir 90-C8 J
Stuart Rd N 157-G4 NN
Stuart Rd S 157-H4 NN
Suburban Pkwy 158-H10 H
Sudbury Way 142-E3 Y
Sue Cir 158-C9 NN
Sue Ct 110-A10 Y
Sugar Bush 88-D1 J
Sugarberry Run 143-J9 H
Sulgrave Ct 105-A1 J
Sulik La 125-D2 Y
Summer Day Ct 157-G1 NN
Summer Dr 141-A7 NN
Summer E 72-J10 J
Summer W 72-J9 J
Summerall Cir 123-H7 F
Summerglen Ridge 140-G4 NN
Summerlake La 140-H5 NN
Summerlins Way 123-H2 NN
Summit Ct 158-A2 H
Sumner Ct 72-C1 J
Sumter Dr 124-B9 NN
Sun Ct 157-G4 NN
Sun Haven Ct 124-C7 NN
Sun Rise Way 141-K2 Y
Sun Valley Ct 124-K6 NN
Sunbriar Way 158-E7 H
Sundance Pl 124-H6 NN
Sundown Ct 141-E8 NN
Sunningdale 88-B2 J
Sunny Meade Cove 141-K7 H
Sunnyside Dr 158-C6 H
Sunnywood Rd 157-G1 NN
Sunrise Ct 124-E9 NN
Sunset Dr 110-B9 Y
Sunset Rd 157-B1 NN
Sunset Rd E 159-D8 H
Sunset Rd W 159-C8 H
Sunset Terr 174-A3 NN
Superior Ct 124-E8 NN
SURA Rd 141-C5 NN
Surrender Rd 108-J6; 109-A7; YE-D6 Y
Surry Ave 158-A10; 174-A1 NN
Surry Ct 159-G3 H
Surry Dr 89-B9 J
Surry Landing Dr 103-E9 SC
Susan Constant Dr 124-B8 NN
Susan Ct 159-A7 H
Susan Dr 126-G7 Pq; 106-F1 Y
Susan Newton La 141-K2 Y
Susquehanna Ct 143-G10 H
Sussex Ct 88-J2 J; 125-D10 Y
Sussex Pl 141-B2 NN
Sutton Pl 157-J3 H
Swain La 142-B3 Y
Swamp Gate Rd 140-E6 NN
Swamp Rd 122-B8 SC
Swan Rd 72-G10 J
Swan Tavern Rise 125-H3 Y
Swann Ave 157-J6 NN
Swanns Pt Cir 159-H2 H
Swanns Pt Rd 103-B10, C10, K9 SC
Swanson Ct 141-F2 Y

Swartmore Dr 142-J10 H
Sweeney Blvd 142-K9 H; 142-K9; 143-A9, D8 L
Sweeney La 160-A6 H
Sweet Gum La 88-A10 J
Sweetbay Arbour 125-D4 Y
Sweetbriar Dr 141-B10 NN
Sweetgum Pl 158-D2 H
Swinley Forest 88-C3 J
Sycamore Ave 174-G2 NN
Sycamore Dr 158-K1 H
Sycamore La 108-D5 Y
Sycamore Landing Rd 41-A9; 57-B1 J
Sycamore Rd 105-A2 J
Sylvia Dr 126-B9 Y
Sylvia La 140-E2 NN
Syms St 159-F6; HE-E3 H
Szetela Ct 158-C5 H

T

Tabatha Cir 157-K1 H
Tabb La 125-J10; 141-J1 Y
Tabb Lakes Dr 141-H2 Y
Tabb St 158-J10 H
Tabbs La 140-D2, E1 NN
Tabernacle Rd 39-H8; 40-A7 NK
Tabiatha La 70-G2 J
Tabor Pk 104-J2 J
Tack Ct 124-E2 NN
Tadich Dr 107-B5 J
Tadworth 88-H3 J
Tahiska La 124-J6 NN
Taliaferro Rd 107-K6 NN
Tall Pine Cir 142-E5 Y
Tall Pines Way 140-K8; 141-A8 NN
Talley Farm Retreat 143-F9 H
Talley Pl 124-A9 NN
Tallpine Dr 158-G5 H
Talltree Pl 158-K1 H
Tallwood Dr 158-E5 H
Tally Farm Retreat 159-G1 H
Tallyho Turn 158-E3 H
Talon Dr 142-C4 Y
Tam-O-Shanter Blvd 106-F1 Y
Tamara Path 157-H1 NN
Tamarisk Quay 158-C1 H
Tameron Ct 124-H6 NN
Tampa Cir 73-E4 Y
Tanbark Dr 157-G1 NN
Tanbark La 56-H7 J
Tandems Way 92-H5 G
Tangier La 140-K2 NN
Tanglewood Cove 105-F1 J
Tanglewood Dr 92-H4 G; 158-J2 H
Tank Farm Rd 91-B7 Y
Tannin Bark Tr 125-E4 Y
Tantallon Dr 107-B5 J
Tanyard St 89-K6; WE-F2 W
Tappan Ave 160-C3 H
Taps Neck Loop 88-G9 J
Tara Ct 72-E9 J; 124-C9 NN
Tarantino Pl 106-G1 Y
Tarleton Bivouac 107-B7 J
Tarpin Cir 57-B6 J
Tarpon Dr 125-J3 Y
Tarrant Rd 142-H10 H
Tarry Pl 157-F1 NN
Tarrytown Ct 157-F1 NN
Tartan La 159-J3 H
Tasker Ct 57-B5 J
Taurus Ct 142-B1 Y
Taxidermy Ct 105-A1 J
Taxiway Dr 175-J7 NBN
Tay River 88-G3 J
Tayloe Cir 106-A4 J
Taylor Ave 123-D5, E5, G7 F; 159-J6 H; 174-D5 NN
Taylor Ave W 143-B9 L
Taylor Dr 73-A7 J
Taylor Farm La 126-A10 Y
Taylor Rd 143-C9 L
Taylor St E 142-K7 H; 143-B9 L
Taylor St W 142-J6 H
Tayside 88-H3 J
Tazewell Rd 124-F8 NN
Tazewells Way 106-B2 J
Teach St 158-J8 H
Teagle La 157-G4 NN
Teague Rd 108-D1 Y
Teakwood Ct 158-E6 H
Teakwood Dr 88-E5 J; 157-H2 NN
Teal Way 72-K10 J
Teardrop La 124-C6 NN
Tecumseh Dr 158-K10 H
Teen Pl 174-E4 NN
Teepee Dr 124-F10 NN
Teleford Dr 140-D1 NN
Telemark Dr 72-G3 J
Telltale Pl 105-H2 J
Temple La 158-A10, B9 NN
Temple St 144-C10 H
Templewood Dr 157-K3 H
Tempsford La 88-J2 J
Tenderfoot Ct 143-H10 H
Tendril Ct 88-K8 J
Tennis Cir 126-G7 Pq
Tennis La 159-J7 H
Teresa Dr 158-A5 H
Terminal Ave 174-B3, C4, D6 NN
Terr Code St 159-A4 H
Terrace 108-C3 Y
Terrace Cir 157-H4 NN
Terrace Dr 141-G10; 157-H1 NN; 126-J10 Pq
Terrace Dr S 126-J10 Pq
Terrace Rd 159-A10 H
Terrebonne Rd 125-C3 Y
Terrell La 158-G2 H
Terrell Rd 140-K9 NN
Terri Beth Pl 140-F5 NN
Terri Sue Pl 124-D3 NN
Terry Ct 158-C2 H
Terry Sue Ct 158-F2 H
Terrys Run 142-F2 Y
Terrywood Ct 125-F2 Y
Tetra Ct 144-B10 H
Tewing Rd 89-B5 J
Tewkesbury Quay 125-G6 Y
Tewkesbury Way 72-K7; 73-A7 J
Texan Dr 142-C4 Y
Thacher Dr 72-D1 J
Thalia Dr 124-B8 NN
Thames Dr 158-B3 H
The Colony 105-A3 J
The Council 89-G5 Y
The Foxes 88-J9 J
The Glebe La 70-A2 CC
The Green 106-A4 J
The Maine 104-C2, C3 J
The Maine W 103-K1; 104-A2 J
The Mews 89-B2 W
The Midlands 89-B8 J
The N Council 89-G5 Y
The Point Dr 70-F2 J
The S Council 89-G5 Y
The Vineyards 105-D3 J
Thea Dr 108-E3 Y
Theatre Rd 141-G1 Y
Theodore St 159-H6 H
Thimble Shoals Blvd 141-D7 NN
Thimble Shoals La 160-C5 H
Thimbleby Dr 124-G5 NN
Thisdell La 158-D10; 174-D1 NN
Thom Hall Dr 159-K5 H
Thomas Athey Ct 157-J2 H
Thomas Bransby 106-A4 J
Thomas Brice 105-K3 J
Thomas Cartwright 106-F4 J
Thomas Ct 73-A10 J
Thomas Dale 106-A3 J
Thomas Dr 127-B10 Pq; 73-H7 Y
Thomas Gates 106-B2 J
Thomas Higgs Ct 55-H5 J
Thomas Landing Rd 92-F3 G
Thomas Nelson Dr 142-B9; 158-C1 H
Thomas Nelson La 89-D2 W
Thomas Rd 125-K1 Y
Thomas Smith La 87-H10 J
Thomas St 159-C5, D5; HE-A1 H
Thompson Cir 123-J7 F
Thompson La 87-K5; 88-A5 J
Thompson St 159-G8 H; 143-E8 L
Thomson La 88-J8 J
Thoreau Cir 142-D5, E5 Y
Thornbriar Ct 158-F7 H
Thorncliff Dr 124-B6 NN
Thornell Ave 143-E8 L
Thornette St 159-C7; HE-A7 H
Thornhill Dr 158-F7 H
Thornrose Dr 125-F5 Y
Thornrose St 159-D3 H
Thoroughbred Dr 142-D7 H
Thorpe Ct 88-F1 J
Thorpes Parish 106-A2 J
Three Notched Rd 125-E4 Y
Three Pt Ct 109-D8 Y
Threechopt Rd 158-A6 H
Thunderbird La 106-G1 Y
Thunderchief Tr 142-C8 Y
Thurgood Ct 159-J4 H
Tidal Dr 141-E8 NN
Tidal Water Rd 39-J1 NK
Tidball Rd 159-K10 FM
Tide Mill La 143-B10; 158-K2; 159-A2 H
Tide Mill Rd 126-A10 Y
Tides Run 109-C9 Y
Tidewater Dr 142-H8 H
Tiffany La 144-C9 H; 124-E3 NN
Tiger Paw Path 141-H3 Y
Tiger Run 141-H2 Y
Tilghman Ct 72-J10; 88-J1 J
Tiller Cir 143-H9 H
Tillerson Dr 124-F10 NN
Tilson Dr 157-G6 NN
Timber Ct 142-E5 Y
Timber La 56-J7 J; 126-G6 Pq
Timber Ridge 88-A10; 104-A1 J
Timberline Cres 141-B10 NN
Timberline Dr 158-B2 H
Timberneck Ct 159-J2 H
Timberneck Dr 92-F1 G
Timberneck Farm Rd 92-D2 G
Timberneck La 141-B1 NN
Timberwood Dr 57-C6 J
Timor Ct 55-D9 J
Timothy Cir 124-A5 NN
Timothy Dell 143-K9 H
Tims La 88-A6 J
Tindall Ct 140-D1 NN
Tindalls Way 158-B1 H
Tinnette Dr 142-B3 Y
Tinsley Ct 124-E6 NN
Tiona Ct 160-A1 H
Tipton Rd 157-A3 NN
Tiverton E 104-B1 J
Tiverton W 104-A1 J
Tivoli Pl 158-C8 NN
Toano Dr 56-B10 J
Toano Woods Dr 72-C1 J
Todd Ct 125-G3 Y
Todd Tr 140-J3 NN
Toddington Cir 89-A8 J
Todds La 157-J4; 158-D4 H
Toddsbury Ct 159-K2 H
Tokay Rd 157-G1 NN
Toledo Ct 142-J10 H
Tolers Rd 106-C1 J
Tom Jones Ct 159-B3 H
Tom Taylor Rd 71-K1 J
Tom Thomas Rd 73-H8 Y
Tomahawk Rd 159-F4 H
Tomahund Dr 86-G9 CC
Tonka Ct 89-E6; WE-A2 W
Topping Cir 90-A9 J
Topping La 158-A4 H
Topsail Ct 158-F4 H
Topsider Ct 124-D3 NN
Torea Ct 72-F3 J
Tormentors La 155-C7 I
Torpedo Al 107-F5 Y
Tour Rd 109-B6; YE-G5 Y
Towels Rd 109-A8 Y
Tower Hill 88-G1 J
Tower La 124-G6 NN; 108-C7 Y
Tower Pl 158-G3 H
Towler Dr 159-A2 H
Towne Centre Way 158-B1 H
Towne Sq Dr 157-H8 NN
Townley Ct 108-G4; YE-B4 Y
Townsend Dr 159-B2 H
Trace N 72-J9 J
Tracy Ct 106-F1 Y
Tracy Pl 123-G7 F
Tradewind Cir 140-C1 NN
Tradewinds Dr 141-E1 Y
Tradewinds Quay 142-G7 H
Trafalgar Ct 90-D7 Y
Trail St 159-E8 H
Trailblazer Blvd 124-F9 NN
Trailer Rd 91-D5 Y
Trails La 124-D3 NN
Tranquility Dr 90-H8 Y
Tratman Ct 142-K10 H
Traverse Rd 141-G5 NN
Travis Cir 160-A2 H; 125-K1 Y
Travis Close 103-F1 J
Travis La 89-A9 J
Travis Pond Rd 103-F1 J
Treasure Ct 124-E6 NN
Treasure Is Rd 105-C4 J
Treasure Key 158-C5 H
Tree Dr 141-E8 NN
Treebark Pl 158-K1 H
Treeclad Pl 158-J1 H
Treefern Pl 158-J1 H
Treeland Ct 124-G7 NN
Treetop Pl 124-A8 NN
Treis Tr 142-E1 Y
Trellis Ct N 124-B6 NN
Trellis Ct S 124-B6 NN
Trent Cir 140-B1 NN
Treyburn Dr 89-D5 J
Triangle Rd 40-C7 NK
Tricia La 125-G2 NN
Trincard Rd 143-G9 H
Trinity Dr 143-E1 Pq; 141-G2 Y
Trinity La 159-F3 H
Triple Crown Ct 142-G6 H
Tripp Terr 157-J1 H
Triton Ct 141-E7 NN
Triton Way 141-J7 H
Trivalon Ct 141-F2 Y
Trolls Path 72-G3 J
Troon 88-G4 J
Trotters Bridge Dr 126-G8 Pq
Trottwood Dr 142-K1 Pq
Trouville Ct 140-J7 NN
Troy Dr 141-A5 NN
Troy Dr E 141-A5 NN
Troy Pl 124-A10 NN
Trudy La 89-B1 J
Truxton St 175-J9 NBN
Truxtun Ct 124-H5 NN
Tubman Dr 89-G6; WE-B2 W
Tuckahoe Dr 140-J7 NN
Tuckahoe St 92-K9 G
Tuckahoe Trace 142-D1 Y
Tucker La 159-F2 H; 141-D8 NN
Tuckers Rd 41-E4 KQ
Tudor Ct 159-A4 H
Tudor Dr 73-E2 Y
Tue Marsh La 110-C6 Y
Tug Boat La 141-D6 NN
Tui Pl 141-K4 Y
Tukaway Ct 141-F8 NN
Tulip Dr 124-D10 NN
Tulip St 160-A6 H
Tully Cove Tr 126-A8 Y
Tupelo Cir 142-B8 H
Turkey Rd 107-C3 J; 107-C3, G2 Y
Turlington Rd 56-C10; 72-B1 J; 140-K8; 141-A7 NN
Turnberry Blvd 124-G9, J8 NN
Turnbridge La 142-E3 Y
Turner Terr 157-K4 H
Turners Landing Rd 54-B6 NK
Turners Neck Rd 71-G5 J
Turning Leaf Dr 55-K1 J
Turpin Pl 140-D1 NN
Tuscon Ct 157-K6 H
Tutters Neck 106-A2, A3 J
Twig La 124-A7 NN
Twin Cr Rd 126-H6 Pq
Twin Lakes Cir 158-F3 H; 123-J9 NN
Twin Oaks Dr 158-D5 H
Two Rivers Rd 87-A10, D10, H10; 103-A1, B1, F1, H1 J
Two Rivers Tr 54-E8 NK
Two Turkey Run 125-D4 Y
Tyburn Ct 143-E10 H
Tyler Ave 123-F4 F; 157-E2, G5 NN
Tyler Brooks Dr 89-C2 W
Tyler Ct 89-H8; WE-D4 W
Tyler Dr 90-C8 J
Tyler Rd 159-F7; HE-E7 H
Tyler St 159-G7; HE-E7, F6 H; 89-H8; WE-D4 W
Tyndal Ct 88-H2 J
Tyndall Cir 159-K3 H
Tyndall Dr 140-G9 NN
Tyndall Pt Ct 92-J9 G
Tyndall Pt Rd 92-J9, K9 G
Tyner Dr 124-F5 NN
Tyree Isle Rd 54-F5 NK
Tysinger Dr 143-E10 H

U

Underwood Rd 106-C2 J
Union St 159-D6; HE-A4 H; 157-H7 NN
University Pl 141-A10, B10; 157-A1 NN
Upland Dr 157-J1 H
Uppershire Way 141-H1 Y
Utility St 107-B10 J

V

Vaiden Dr 89-A9 J
Vaidens Ct 38-C9 NK
Vaidens Pond Rd 38-B10 NK
Vaior Ct 90-C5 Y
Valasia Rd E 126-H10 Pq
Valasia Rd N 126-H10 Pq
Valasia Rd S 126-G10 Pq
Valentine Ct 141-J10; 157-J1 H; 156-H1 NN; 125-H8 Y
Valirey Dr 159-F1 H
Valley Forge Dr 141-B3 NN
Valley Front La 58-K7 G
Valmoore Dr 126-F8 Pq
Valor Pl 159-E4 H
Van Dyke La 140-J9 NN
Van Patten Dr 159-D9 H
Van Tol Rd 90-H5 Y
Van Voorhis St 123-H7 F
Vanasse Ct 158-G2 H
Vantage Ct 141-D2 NN
Vantage Dr 126-E10 Pq
Vass La 89-A7 J
Vaughan Ave 158-G9, G10 H
Veneris Ct 143-F10 H
Venetia Ct 124-B10 NN
Ventnor Dr 124-A10 NN
Ventura Way 124-G7 NN
Venture La 88-H10 J
Venture Way 144-D9 H
Vera Cir 141-G7 NN
Verde Quay 158-C1 H
Verell St 158-K9 H
Verline Ct 124-B8 NN
Vernon Cir 158-E3 H
Vernon Pl 158-A7 NN
Vernon St 92-K10; 108-K1 G
Vicky Ct 140-D2 NN
Victor St 159-C8 H
Victoria Blvd 158-J10; 159-A9, D7; 174-G1 H
Victoria Blvd E HE-C7 H
Victoria Blvd W HE-A8 H
Victoria Ct 107-B5 J
Victoria St 92-H2 G
Victoria Sta 124-G6 NN
Victory Ave 174-A4 NN
Victory Blvd 126-E10; 142-G1 Pq; 126-E10; 141-F2, J1 Y
Victory Rd 125-J5 Y
Victory St 142-J6 H
Vienna Cir 142-A7 H
View Pointe Dr 107-G8 NN
View Rd E 92-H3 G
Viking Dr 140-B3 NN
Viking Rd 72-G3 J
Viking Way 142-G5 H
Villa Rd 157-E5 NN
Villa Way 125-H10 Y
Village Ave 141-F3 Y
Village Dr 92-K8 G
Village Green Pkwy 141-C3, C4 NN
Village Landing Dr 92-F2 G
Village Pk Dr 104-H1 J
Village Pk Dr E 88-J10; 104-H1 J
Village Pk Dr W 88-H10 J
Village Pkwy 158-B1 H; 157-G5 NN
Vincent Dr 72-C1 J
Vine Dr 125-J8 Y
Vine St 159-B8 H
Vineyard La 90-D8 Y
Vintage Ct 88-G8 J
Violet St 160-A6 H
Virginia Ave 159-J6 H; 175-J8, J10 NBN; 89-F7; WE-B3 W
Virginia Dr 140-A1 NN
Virginia La 125-F6 Y
Viscount Dr 124-K9 NN
Vista Ct 89-H10 W
Vista Dr 124-B10 NN
Vivian Cir 140-E3 NN
Vivian Ct 108-A4 Y
Von Schilling Dr 158-H3 H
Von Steuben Dr 107-J6 NN
Voodoo Dr 142-C8 Y
Voyager Dr 142-G4 H
Vreeland Dr 126-E3 Y

W

Waco Ct 143-K10; 159-K1 H
Wade Cir 140-F2 NN
Wade Curtis Rd 107-C4 J
Wade Rd 144-B9 H
Wagner Rd 126-H6 Pq
Wahrani La 38-K9; 54-J1 NK
Wainwright Dr 126-H9 Pq
Wainwright Dr W 126-G9 Pq
Wainwrights Bend 125-E5 Y
Wake Dr 89-E7; WE-A4 W
Wakefield Ave 158-H10 H
Wakefield Rd 141-C9 NN
Wakerobin Rd 105-G2 J
Wal Mart Way 140-J1 NN
Walcott St E 142-K6 H
Walcott St W 142-K6 H
Walden Dr 92-F4 G; 126-A7 Y
Walden Pond Ct 124-D8 NN
Wales Cir 158-C2 H
Wales Ct 105-J2 J
Walker Dr 72-E9 J
Walker Rd 107-F4 Y
Walker Rd E 158-E5 H
Walker Rd N 158-E5 H
Walker Rd W 158-E5 H
Walker St 123-H5 F
Walkers Landing La 143-E10 H
Walkin La 109-K10; 125-J1 Y
Wallace Cir 140-J10 NN
Wallace Ct 141-K2 Y
Wallace Rd 144-B9 H; 90-D10 J
Waller Mill Rd 89-F5, G3 Y
Waller St 89-K7; WE-E3 W
Walnut Ave 143-B9 L; 174-G3 NN
Walnut Dr 108-F4; YE-A4 Y
Walnut Grove 140-J9 NN
Walnut Hills Cir 89-D9 W
Walnut Hills Dr 89-D9 W
Walnut St 159-E2 H
Walters La 126-A2 Y
Walters Rd 140-D6 NN
Waltham Cir 123-K10 NN
Waltham St 142-J10; 158-J1 H
Walton Heath 88-G3 J
Walton La 159-B8 H
Walton St 92-K10 G
Waltons Approach 142-E2 Y
Waltrip La 105-D2 J
Waltz Farm Dr 89-C2 W
Wanger Cir 140-F4 NN
Warbler Pl 105-B2 J
Ward Ct 141-B1 NN
Ward Dr 158-H7 H
Ward Rd 143-D5, F7 L
Ware Cr Rd 56-E3, H6 J
Ware La 72-H4 J; 140-J2 NN
Ware Rd 104-K3 J
Warehams Pond Rd 106-E3, F5, G4 J
Warehams Pt 106-E4 J
Warehouse Course 125-G2 Y
Warehouse Rd 142-J9 H
Warhawk Dr 142-C8 Y
Warhill Tr 72-G10 J
Warner Hall Pl 124-H5 NN
Warner Rd 158-A4 H
Warner St S 142-J6 H
Warren Dr 124-A7 NN
Warren La 39-K1 NK
Warren St 159-J5 H
Warrens Pond Rd 71-K1 J
Warrington Cir 159-J2 H
Wartons Way 143-J9 H
Warwick Blvd 107-F7; 123-J2; 124-A4, D9; 140-F1, J5; 141-A7; 157-C2, J8; 173-K1; 174-A2 NN
Warwick Cres 157-E5 NN
Warwick Ct 90-E9 Y
Warwick Hills 88-G5 J
Warwick Landing Pkwy 124-A5 NN
Warwick Springs Dr 140-H1 NN
Warwickshire Ct 157-G5, H5 NN
Washington Ave 173-K2; 174-A3 NN
Washington Blvd 123-F7, G6 F
Washington Ct 142-A5 Y
Washington Dr 107-H7 NN
Washington Rd 109-B5 Y
Washington Sq Dr 125-G6 Y
Washington St 108-K1 G; 159-F5; HE-E2 H; 89-K6; WE-E2 W
Water Country Pkwy 90-H9 Y
Water Fowl Dr 126-A3 Y
Water Oak Ct 141-C1 NN
Water St 159-K8 FM; 108-H2, J3; YE-A1, D2 Y
Waterford Cir 157-J3 H
Waterford Ct 88-J2 J
Waterfowl La 140-F7 NN
Waterfront Cir 174-E5 NN
Watergate Terr 140-F10 NN
Waterman Dr 141-A4 NN
Watermans Way 109-C9 Y
Watermill Run 140-K5 NN
Waters Edge Cir 143-D10 H
Waters Edge Dr 88-G8 J; 141-C7 NN
Waters Ridge Rd 125-H5 Y
Waterside Dr N 38-B6; 54-D1 NK
Waterside Dr S 54-C6, D3 NK
Waterside Pl 125-H5 Y
Waterton 88-D2 J
Waterview Dr 123-J10 NN
Waterview Pt 142-A7 H
Waterview Rd 71-K2 J; 109-F7, H6 Y
Watford La 89-C6 J
Watkins Cir 143-B1 Pq
Watkins Dr 158-H7 H
Watson Dr 141-D2 NN
Watson La 160-B1 H
Watts Ave 143-C5 L
Watts Dr 142-H10; 158-H1 H
Waverly La 55-C9 J
Waverly Pl 124-H5 NN
Waverly St 159-D6; HE-B3 H
Wax Myrtle Dr 90-A6; WE-F2 W
Wayanoke Ct 159-H2 H
Wayfin Cir 140-H6 NN
Weathers Field Way 72-K7; 73-A8 J
Weaver Rd 90-F10; 106-F1 Y
Weaver Rd E 158-F5 H
Weaver Rd W 158-C6 H
Weber La 159-K6 H
Webster St 159-J7 H
Wedgewood Ct 39-B8 NK
Wedgewood Dr 159-D2 H; 39-C8 NK; 141-K7 NN
Wedgewood Dr E 125-D10 Y
Wedgewood Dr N 141-K6 NN
Wedgewood Dr W 125-C10 Y
Wedgewood La 39-B8 NK
Weldon Pl 55-K5 J
Weldun Ct 123-K8 NN
Welford La 141-B9 NN
Wellesley Blvd 73-A8 J
Wellesley Dr 140-K8; 141-A8 NN
Wellie Ct 124-E10 NN
Wellington Cir 104-F1, F2 J; 107-J10 NN
Wellington Dr 157-K4 H; 90-D7 Y
Wells Ct 158-F2 H
Wells Rd 140-C1 NN
Welstead St 56-C5 J
Wendel Dr 142-J10 H
Wendell Cir 126-G7 Pq
Wendfield Cir 141-G5 NN
Wendwood Dr 140-J4 NN
Wendy Ct 141-H10 NN
Wenley Cir 155-B6 I
Wentworth 88-H3 J
Wentworth Pl 157-J3 H
Wesley Ct 125-J4 Y
Wesley Pond Ct 124-D8 NN
Wesleyan Ct 125-B10 NN
Wessex Hundred Rd 105-D3 J
West Ames St 142-H5 H
West Autumn 72-H9 J
West Ave 174-A3 NN
West Bay 104-B1 J
West Bayberry Ct 159-E2 H
West Beechwood Dr 107-D6 J
West Boundary Rd 107-A4, E6 J; 107-E6 NN; 106-J2; 107-E6 Y
West Bristol La 142-D4 Y
West Brittington 104-B1 J
West Bush Rd 142-H4 H
West Cemetery La 143-E3 Pq
West Chester 88-C2 J
West Cir 104-A2 J
West Co St 159-H7 H
West Dressage Ct 142-D7 H
West Durand St 142-J6 H
West Francis Chapman 104-G1 J
West Governor Dr 140-D6 NN
West Is Rd 103-F1 J
West John Proctor 104-G1 J
West Kingswood Dr 105-A1 J
West La 89-F8 W
West Lamington Rd 159-D1 H
West Landing 106-A5 J
West Lewis Dr 158-C7 H
West Links 88-E3 J
West Little Back River Rd 159-E2 H
West Manor Dr 90-D8 Y
West Mercury Blvd 157-J8; 158-C5, H4; 159-A3 H; 157-J8 NN
West Miller Rd 39-G9 J
West Pembroke Ave 158-G9; 159-A7; 174-E1; HE-A3, B3 H
West Pinto Ct 142-D6 H
West Preston St 159-D1 H
West Queen St 158-G5, H6 H
West Queens Ct 159-E6; HE-D5 H
West Queens Dr 90-D4 Y
West Queens Way 159-E6; HE-C5 H
West Rd 91-J9; 92-A10 Y
West Reid St 142-J6 H
West Rexford Dr 123-K6 NN
West Richmond Rd 54-H3; 55-A4, F6 J
West River La 126-D4 Y
West River Pt Dr 143-H9 H
West River Rd 126-G8 Pq
West Riverside Dr 54-A7 NK
West Roan Ct 142-D6 H
West Royal Norfolk 88-H3 J
West Russell Ct 158-B10 NN
West Sand Dr 56-D6 J
West Sandy Pt Rd 126-J7 Pq
West Savannah Sq 89-E4 Y
West Semple Rd 90-C8 Y
West Settlers Landing Rd HE-A5 H
West Southampton Ave 159-C8 H
West Spring 72-H10 J
West St 159-C4 H
West Steeplechase Way 88-H8 J
West Stoney Cr 104-B1 J
West Summer 72-J9 J
West Sunset Rd 159-C8 H
West Taylor Ave 143-B9 L
West The Maine 103-K1; 104-A2 J
West Tiverton 104-A1 J
West Victoria Blvd HE-A8 H
West Vill Pk Dr 88-H10 J
West Wainwright Dr 126-G9 Pq
West Walcott St 142-K6 H
West Walker Rd 158-E5 H
West Weaver Rd 158-C6 H
West Wedgewood Dr 125-C10 Y
West Whittaker Close 103-G1 J
West Woodland Dr 125-K8 Y
West York Haven Dr 58-J10 G
Westbriar Dr 158-F7 H
Westbrook Dr 158-J3; 159-A3 H
Westbury 88-H5 J
Westcreek Ct 157-H3 H
Western Dr 124-E9 NN
Westfield Pl 158-H10 H
Westgate Cir 89-C1 W
Westgate Ct 141-C1 NN
Westlake Ct 107-C5 J
Westlawn Dr 144-C10 H
Westly Ct 159-G3 H
Westminister Dr 159-A4 H
Westminster Pl 73-E2 Y
Westmont Dr 142-A6 H
Westmoreland Dr 159-F3 H; 72-K10; 73-A10; 88-K1 J
Westmoreland La 58-K1 G
Weston Ct 124-G5 NN; 89-E10 W
Weston Rd 126-B5 Y

Westover Ave 89-F6; WE-A1 W
Westover Dr 126-E9 Pq; 90-E9 Y
Westover Dr N 126-E9 Pq
Westover Rd 157-E5 NN
Westover St 159-D2 H
Westpark La 142-B9 H
Westphal Dr 159-F1 H
Westport Cres 125-C10; 141-B1 NN
Westview Dr 141-K7 H
Westward Ho 88-C3 J
Westwind Dr 124-K9 NN
Westwood Ave 159-A7 H
Wetherburn La 88-K2 J
Wethersfield Pk 125-G6 Y
Wexford Ct 141-K5 Y
Wexford Hill Rd 157-J2 H
Wexford Run 104-B1 J
Weyanoke La 124-G5 NN
Weyland Rd 143-A6 L
Weymouth Dr 141-A1 NN
Weymouth Terr 158-B3 H
Whaler Dr 124-D3 NN
Wharf Row 125-G3 Y
Wharfview Ct 92-F2 G
Whealton Rd 158-A5, A6 H
Wheatland Dr 158-K2; 159-A3 H
Wheeler Ave 174-G1 H
Wheeler Dr 123-J9 NN
Wheely Cir 125-C2 Y
Whetstone Dr 157-K3 H
Whisley Ct 124-H4 NN
Whispering Pine Dr 125-J4 Y
Whispering Way 126-A6 Y
Whisperwood Dr 140-B1 NN
Whistle Walk 88-K8 J
Whistler La 160-A10 FM
Whitaker Ave 160-B4 H
Whitaker Ct 88-G2 J
Whitby Ct 89-D10 W
Whitby Mews 89-B2 J
White Acre Rd 57-D6 J
White Cedar Ct 142-B8 H
White Ct 88-G1 J
White Hall Cir 159-J2 H
White House Cove 141-B1 NN
White House St HE-G3 H
White Marsh Ct 159-H2 H
White Oak Dr 54-H10 J; 54-A7 NK; 157-D3 NN
White Oak Tr 159-F2 H
White Stone Ct 107-H9 NN
Whitebrook La 140-G4 NN
Whitehall Ct 88-H1 J
Whitehaven Cir 89-A8 J
Whitehorn Cir 89-A8 J
Whitehouse Cir 126-H9 Pq
Whitehouse Dr 126-K8; 127-A8 Pq
Whitehouse Landing La 92-F1 G
Whitehouse St 159-G6 H
Whiteridge La 141-K4 Y
Whites La 141-A8 NN; 143-E1 Pq; 109-K8 Y
Whites Rd 125-C2 Y
Whitewater Dr 124-C3 NN
Whiting Ave 107-A4 J
Whiting St 159-E7; HE-C6 H
Whitman Pl 160-A4 H
Whitman Rd 140-F3 NN
Whitney Ct 144-A10; 160-A1 H
Whits Ct 140-H8 NN
Whitt Ct 108-G4; YE-A4 Y
Whittaker Close E 103-H1 J
Whittaker Close W 103-G1 J
Whittaker Is Rd 103-G1 J
Whittakers Mill Rd 106-C2 J
Whittier Ave 141-A8 NN
Whittles Wood Rd 103-C1 J
Wichita La 73-E4 Y
Wicked Ct 157-G4 NN
Wickham Ave 158-A7 H; 158-A7, B8; 174-D2, E4 NN
Wickham La 144-B9 H
Wickham Pl 174-E4 NN
Wickre St 90-B8 J
Wicomico La 140-J2 NN
Wicomico Turn 142-C2 Y
Wicomico Vill Dr 92-H4 G
Widgeon Cir 140-B2 NN
Wigner Ct 160-B3 H
Wilcox La 157-J6 NN
Wild Duck Ct 141-K8 H
Wilderness La 72-J1 J
Wilderness Rd 143-G10 H
Wilderness Way 124-F4 NN
Wildey Rd 110-E10; 126-E1 Y
Wildflower Ct 88-H9 J
Wildwood Ct 140-C3 NN
Wildwood Dr 158-D8 H; 126-A7 Y
Wilhelmina St 89-C5 W
Wilkins Dr 90-D8 Y
Wilkinson Dr 73-F7 Y
Will Scarlet La 90-D4 Y
Will Scarlet La N 90-D4 Y
Willard Ave 159-J8 H
Willard Pl 140-G7 NN
Willards Way 142-E1 Y
Willbrook Rd 141-D3 NN
Willcox Neck Rd 54-C10; 70-A10, B8, C3; 86-A3, A4 CC
William Allen 105-K3 J
William Barksdale 105-K4 J
William Bedford 104-G2 J
William Carter Rd 90-E9 Y
William Claiborne 106-B3 J
William Hodgson 88-B7 J
William Lee 88-B7 J
William Richmond 105-K3 J
William S 106-A3 J
William Way 89-F10, G9, G10 W
Williams Cir 73-A7 J; 126-C5 Y
Williams Landing Rd 92-F3, G4 G
Williams St 159-J8 H; 174-E2 NN
Williamsburg Ave 89-H8; WE-D4 W
Williamsburg Ct 140-J8 NN
Williamsburg Dr 158-H2 H
Williamsburg Glade 104-J4 J
Williamsburg Landing Dr 105-D1 J
Williamsburg Plantation Dr 88-K2; 89-A2 J
Williamsburg Pottery Rd 72-J5 J; 72-J5; 73-A5 Y
Williamsburg West Dr 88-J3 J
Williamson Dr 72-D9 J; 124-A4 NN
Williamson Pk Dr 124-A4 NN
Williford St 123-J7 F
Willis Ch Yard 159-F2 H
Willis Ct 142-G2 Pq
Willis Dr 140-H7 NN
Willnew Dr 159-C5 H
Willoughby Dr 90-E4 Y
Willoughby Pl 159-A9 H
Willow Bend Ct 124-D8 NN
Willow Cove 141-C1 NN
Willow Ct N 124-B5 NN
Willow Ct S 124-B5 NN
Willow Dr 105-D1 J; 157-G3 NN
Willow Green Dr 141-F4 NN
Willow La 92-H2 G; 56-H7 J
Willow Leaf Dr 126-C2 Y
Willow Oaks Blvd 143-H10; 159-H1 H
Willow Pt 141-C1 NN
Willow Rd 144-C8 H
Willow Springs Ct 105-E1 J
Willow St 143-B9 L
Willowood Dr 142-D4 Y
Willows Pl 158-K1 H
Willowtree Rd 158-E4 H
Wills Cir 125-K1 Y
Wills Way 157-J1 H
Wilmont La 123-K10 NN
Wilson Ave 123-E6, F7, J6, J7 F
Wilson Cir 58-K5 G; 72-G2 J; 141-A10 NN
Wilson Dr 126-H7 Pq; 73-J4 Y
Wilson Farm La 126-C10 Y
Wilson La 159-K6 H
Wilton Ave 160-B5 H
Wiltshire Cres 124-H5 NN
Wiltshire Pl 160-B5 H
Wimbledon Terr 158-C3 H
Wimbledon Way 88-H9 J
Winchester Dr 158-D2 H
Winchester Rd 90-F7 Y
Wind Forest La 125-J5 Y
Wind Mill Pt La 144-B6 H
Windbrook Cir 141-B2 NN
Windbrook La 89-D5 J
Windemere Rd 140-B2 NN
Winder Cres 140-K8 NN
Winder Ct 158-D8 H
Winder Farm La 142-G4 H
Winder Rd 125-H10 Y
Winders La 125-H5 Y
Windjammer Cres 140-C2 NN
Windjammer Ct 107-B5 J
Windjammer Dr 143-H9 H
Windmere La 89-F6; WE-B1 Y
Windsong La 142-D5 Y
Windsor Castle Rd 124-G7 NN
Windsor Ct 124-E5 NN
Windsor Dr 158-K4 H
Windsor La 89-F6; WE-B1 Y
Windsor Pines Way 124-G7 NN
Windsor Pl La 89-F5 Y
Windsor Way 88-J2 J
Windstar 88-E1 J
Windy Knoll Rd 40-A7 NK
Windy La 39-K7 NK
Windy Pt Rd 142-K1 Pq
Windy Ridge La 140-G4 NN
Windy Shore Dr 142-E6 Y
Windy Tree La 123-K6 NN
Wine St 159-E5, F6; HE-D2, D4 H
Winfred Rd 92-J5 G
Winfree Dr 160-A2 H
Winfree La 125-H2 Y
Wingate Dr 89-G5; WE-B1 Y
Winged Foot 88-G6 J
Wingfield Close 87-H10 J
Wingfield Dr 158-E7, F7 H
Wingfield Lake Rd 87-H10 J
Winggapo Dr 54-C6 NK
Winnard Rd 159-J1 H
Winona Dr E 158-K9 H
Winona Dr N 158-K9 H
Winslow Dr 124-E3 NN
Winsome Facts 106-D3 J
Winsome Haven Dr 125-J1 Y
Winston Ave 141-D10 NN
Winston Dr 89-B10 J
Winston Pl 141-E10 NN
Winter Ct 141-J1 Y
Winter E 72-J9 J
Winterhaven Dr 141-A7 NN
Winterset Pass 73-C8 J
Winthrop Rd 90-B8 Y
Winthrop Terr 158-F5 H
Winthrope Dr 125-B10; 141-B1 NN
Wise Dr 89-C4 J
Wise Rd 160-A3 H
Wisteria Garden Dr 107-A6 J
Witch Hazel Dr 73-G8 Y
Witch Hazel Mill Rd 126-D10 Y
Witness La 124-D8 NN
Wolf Cr 88-D4 J
Wolf Dr 141-D9 NN
Wolf Rd 107-G3, J3 Y
Wolf Trap Rd 109-F10; 125-G1 Y
Wolftrap 158-B7 NN
Woltz Ct 159-G1 H
Womack Dr 159-J3 H
Womble Ct 160-B3 H
Wood Ave 160-B3 H
Wood Duck La 140-F6 NN
Wood Pheasant Run 72-J9 J
Wood Pond Cir 104-K2 J
Wood Post Ct 124-D8 NN
Wood Row 123-E7 F
Wood Violet La 88-J9 J
Woodall Ct 124-E6 NN
Woodall Dr 142-F7 H
Woodbine Ct 105-H1 J
Woodbine Dr 105-H1 J
Woodbridge Dr 159-B3 H; 124-C3 NN
Woodbrook La 92-G2 G
Woodbrook Run 140-J5 NN
Woodburn Dr 160-C1 H
Woodbury Ct 140-D4 NN
Woodbury Forrest Dr 158-D5 H
Woodcreek Ct 158-D2 H
Woodcreek Dr 124-D3 NN
Woodcrest Dr 159-J5 H
Wooded Hill Dr 160-B1 H
Woodfin Rd 157-G7, H6 NN
Woodfin St 159-G8 H
Woodford Ave 92-H5 G
Woodhall Spa 88-J3 J
Woodhaven Dr 142-G3 Pq; 126-B7 Y
Woodhaven Rd 124-A6 NN
Woodlake Cir 140-G10 NN
Woodlake La 158-A1 H
Woodlake Run 125-H6 Y
Woodland Dr 157-B3 NN; 126-J9 Pq
Woodland Dr E 125-K8 Y
Woodland Dr W 125-K8 Y
Woodland Rd 159-H6, J3 H; 56-K7; 57-A7 J; 125-K8 Y
Woodlawn Dr 157-K3 H
Woodmansee Dr 160-A2 H
Woodmere Ct 89-D10 W
Woodmere Dr 89-D10 W
Woodmont La 72-E2 J
Woodnote La 124-G4 NN
Woodpath La 158-A2 H
Woodridge Ct 72-B1 J
Woodroof Rd 140-J10 NN
Woodrose Pl 158-J1 H
Woodrum Pl 126-F8 Pq
Woods Dr 89-K5; WE-F1 W
Woods La 158-A1 H
Woods Rd 141-E8, F10; 157-F1 NN; 110-A9 Y
Woods Walk Ct 104-B1 J
Woodsedge La 141-J2 Y
Woodside Dr 159-H3 H; 107-A5 J
Woodside La 124-F3, F5, H5, J7 NN
Woodside St 92-H2 G
Woodsman Dr 158-C2 H
Woodsville Rd 92-G2 G
Woodview La 158-C1 H
Woody Cir 160-A2 H
Woolridge Pl 157-D3 NN
Worcester Way 141-B2 NN
Worden Ave 159-E9 H
Workington 88-B2 J
Workshop 88-E1 J
Worley Rd 142-K4; 143-A5, B5 L
Wormley Cr Dr 109-C8, D7 Y
Wormley Cr Rd 109-D6 Y
Wornom Dr 110-A9 Y
Worplesdon 88-D2 J
Worster St 159-B7 H
Wreck Shoal Dr 141-D7 NN
Wrenfield Dr 57-F9 J
Wrenn Cir 124-E7 NN
Wrexham Ct 159-J2 H
Wright Ave 143-D8 L
Wright Cir 108-E1 Y
Wright Dr 158-H6 H; 157-K9 NN
Wright St 123-H4 F; 142-J5 L
Wright St S 142-J6 H
Wrights Dock Rd 41-E5 KQ
Wriothesley St 159-E7; HE-C8 H
Wrong Rd 155-F6 I
Wrought Iron Bend 141-K5 Y
Wyatt Cir 105-J3 J
Wyatt St 159-E2 H
Wylewood Pl 72-E2 J
Wyman Ct 38-C10 NK
Wyn Dr 124-B5 NN
Wyndham Ct 124-H4 NN
Wyndham Dr 158-G2 H
Wyndham Way 88-J1 J
Wynne Rd 126-C2 Y
Wynstone Ct 141-D2 NN
Wynterset Cir 125-H2 Y
Wyse Ct 142-J7 H
Wythe Ave 89-F6; WE-A2 W
Wythe Cr Rd 142-G4 H; 126-H10; 142-H3 Pq
Wythe Cresent Dr 174-K1 H
Wythe Ct 142-A5 Y
Wythe Hgts 159-C9 H
Wythe La 89-G8; WE-B4 W
Wythe Landing Loop 142-H4 H
Wythe Pkwy 158-K10; 174-K1 H

Y

Yale Dr 158-C6 H
Yarmouth Cir 140-J4 NN
Yarrow Ct 88-J10 J
Yates Dr 89-F7; WE-A3 W
Yeardley Dr 157-D3 NN
Yeardleys Grant 106-C3 J
Yearling Ct 142-E4 Y
Yeaton Dr 109-D5 Y
Yellowwood Rd 158-K1 H
Yoder La 140-E3 NN
York Cir 158-A10 NN
York Crossing Rd 125-G6 Y
York Downs Dr 142-E4 Y
York Dr 108-C5 Y
York Haven Dr W 58-J10 G
York Haven La 58-K9; 74-K1 G
York La 92-K7 G; 108-K9; 109-A9 Y
York Pt Dr 126-H2 Y
York Pt Rd 110-F10; 126-G1 Y
York Rd 108-B2 Y
York River Dr 92-K9 G
York River La 140-J2 NN
York River Rd 41-E1, F2 KQ
York River State Pk Rd 57-D7 J
York Shores Dr 92-K9 G
York St 159-A8 H; 89-K7; WE-E3, F4 W
York Warwick Dr 108-J9; 109-A9 Y
Yorkshire Dr 89-D10 W; 125-G10 Y
Yorkshire La 124-G5 NN
Yorkshire Terr 158-B3 H
Yorktown Rd 107-J7, J8 NN; 125-J10; 142-A1, B1 Y
Yorktown Rd E 126-E10 Pq
Yorkview Rd 109-D7 Y
Yorkville Rd 125-J6; 126-A6 Y
Yorkwood La 125-H6 Y
Youngs Mill La 140-H3 NN
Youngs Rd 140-E5 NN
Yukon St 159-K7 H
Yulee Ct 143-K9 H
Yves Cir 157-H4 NN

Z

Zachary Pl 142-E1 Y
Zefstat La 107-A10 J
Zekova Reach 158-C1 H
Zelkova Rd 90-A6; WE-F2 W
Zilber Ct 143-K9 H
Zinzer Rd 160-B2 H
Zipp Dr 158-A10 NN
Zoeller Ct 141-C2 NN
Zweybrucken Rd 108-J4; YE-D4 Y

NUMBERED STREETS

1st Ave 142-C5 Y
1st Light Ct 124-H8 NN
1st Patent Ct 88-F7 J
1st St 160-C5, D4, E2 H; 90-C10 J; 175-J10 NBN; 89-J6; WE-E2 W; 91-B6; 109-K6; 125-C2 Y
2nd Ave 142-B5 Y
2nd St 160-C4, D4 H; 175-H10 NBN; 90-A7; WE-F3 W; 91-B5; 109-K7; 125-C2 Y
3rd Ave 142-B5 Y
3rd St 158-H10 H; 175-H10 NBN; 91-C5 Y
3th St 109-J7 Y
4th Ave 142-B5 Y
4th St 160-C4 H; 91-D5; 109-J7 Y
5th Ave 90-B9 J; 142-B5 Y
5th St 123-J6 F; 160-C4 H; 174-F1 NN; 109-H7 Y
6th St 123-H6 F; 160-C4 H; 174-E6 NN; 109-H7; 142-C5 Y
7th Ave 142-C5 Y
7th St 123-H6 F; 160-C3 H; 109-H7 Y
8th Ave 142-C5 Y
8th St 123-H6 F; 160-B3 H; 124-J9 NN; 109-H8 Y
9th Ave 142-C5 Y
9th St 109-G7 Y
10th Ave 142-C5 Y
10th St 123-H6 F; 109-G7 Y
11th St 123-H6 F; 174-E5 NN; 109-G8; 142-C5 Y
12th St 123-H6 F; 174-D5, E5 NN; 142-A4 Y
13th St 123-G6 F; 174-D5, E4 NN; 142-A5 Y
14th St 123-G6 F; 174-C5, D4, E4 NN; 142-A4 Y
15th St 174-E4 NN
16th St 174-D4 NN
17th St 174-C5, D4, E4, F4 NN
18th St 123-G4 F; 174-B5, D4, E4, F3 NN
19th St 123-F4 F; 174-C4, E4, G2 NN
20th St 174-C4, E3 NN
21st St 174-C4, E3 NN
22nd St 174-C4, D4, F3 NN
23rd St 174-A4, E3 NN
24th St 123-F6 F; 174-A4, D3, E3 NN
25th St 123-J5 F; 174-A4, D3, G2 NN
26th St 123-H5 F; 174-A4, C3 NN
27th St 174-A4, C3 NN
28th St 123-H4 F; 174-A4, C3 NN
29th St 174-A4, C3, D2, F2, G1 NN
30th St 174-A4, C3, F2 NN
31st St 174-A4, B3, D2 NN
32nd St 174-A3, B3, D2 NN
33rd St 174-A3, B3, D2 NN
34th St 174-A3, C2, D2 NN
35th St 174-A3, C2, E1 NN
36th St 174-A3, C2, E1 NN
37th St 174-A3, D2, E1 NN
38th St 174-A3, D1 NN
39th St 174-A3, C2 NN
40th St 174-B2 NN
41st St 174-A2, B2, D1 NN
42nd St 174-A2, B2, D1 NN
43rd St 174-A2, B2, D1 NN
44th St 173-K2; 174-A2, B1, C1 NN
45th St 173-K2; 174-B1 NN
46th St 173-K2; 174-A2, B1 NN
47th St 173-K2; 174-B1 NN
48th St 158-C10; 173-K2; 174-B1 NN
49th St 173-K2; 174-B1 NN
50th St 158-F9, G9 H; 173-K1; 174-B1 NN
51st St 173-K1 NN
52nd St 158-F9 H; 173-K1 NN
53rd St 173-K1 NN
54th St 158-C10; 173-K1 NN
55th St 173-K1 NN
56th St 158-F9 H; 158-D9; 173-K1 NN
57th St 173-K1 NN
58th St 158-E9 H; 157-K10; 158-A10; 173-K1 NN
59th St 157-K10 NN
60th St 158-F8 H; 157-K10; 158-C9 NN
61st St 157-K10 NN
62nd St 157-K10 NN
63rd St 157-K10 NN
64th St 157-J10 NN
65th St 157-J10 NN
66th St 157-J10 NN
67th St 157-J9 NN
68th St 157-J9 NN
69th St 157-J9 NN
70th St 157-J9 NN
71st St 157-J9 NN
72nd St 157-J9, K9; 158-B8 NN
73rd St 157-J9, K8; 158-B8 NN
74th St 157-K8; 158-B8 NN
75th St 157-H8; 158-B8 NN
76th St 158-B7 NN
77th St 158-A8 NN
78th St 158-A7 NN
79th St 157-K8; 158-A7 NN
80th St 158-A7 NN
81st St 158-A7 H
82nd St 157-K7; 158-A7 NN
105th St 175-J8 NBN
107th St 175-J8 NBN

ROUTE NUMBERS

I 64 141-K9; 158-C1, F3; 159-A5, G6, J10; 175-K1; HE-B1, G5, H7 H; 39-C9; 55-E1; 56-A4, J10; 72-K1; 73-A2; 107-A4 J; 38-A3, J7; 39-C9 NK; 107-G7; 108-A10; 124-A1, D6, G10; 140-H1; 141-B2 NN; 73-E6, K10; 90-B3, H10; 106-H1 Y
I 664 158-E10, J7 H; 174-C1, F10 NN
US 17 92-H1, K6, K10; 108-K1 G; 173-B3 I; 141-F8, F10, G5; 157-F1, G9, H5; 173-B3 NN; 108-K8; 109-A9; 125-C1, F4, G9; 141-G1; YE-C2 Y
US 60 158-K10; 159-A10, G6, J10; 174-H1; 175-K1; HE-B8, C6, H7 H; 54-K4; 55-A4, J6; 56-B8, D10; 72-F1, J5; 73-A7, C10; 89-C1; 90-D10; 106-E1, K4; 107-A6, E7 J; 54-A3 NK; 107-J10; 123-J1; 124-B6, E10; 140-F1, K6; 141-A7, B10; 157-C1, E4, H7, K10; 173-K1; 174-A2, C3 NN; 89-E4, J5; WE-D1, F3, F4 W; 106-E1, K4 Y
US 143 90-A3 Y
US 258 159-K9 FM; 158-B6, F4, K3; 159-A4, H5 H; 157-G9 NN
VA 5 86-A4, E6, J9 CC; 87-A8, F10; 88-A10, E10, K10; 89-A10; 104-F1 J; 89-E9, H6, K5; WE-B4, C3, D2, F1, F3 W
VA 30 39-F9; 55-G1, J4; 56-G8 J; 39-A1, F7 NK
VA 31 89-C10; 104-D5, K3; 105-C1 J; 89-E8 W
VA 32 173-B3 I; 157-G9; 173-B3 NN
VA 33 38-D3, J1 NK
VA 105 123-H4; 124-C2 NN; 109-B10; 124-H2; 125-A1 Y
VA 132 89-G10, K6; 105-H1; WE-C2, C3, C4 W; 89-K4; WE-D1 Y
VA 132Y 89-H6; WE-D2 W
VA 134 142-G6, G10; 158-F1, F3, K4; 159-A4, D6; HE-C4 H; 141-H1; 142-B4 Y
VA 143 159-K9 FM; 158-K9; 159-B8, G6; 174-H1; HE-B7, C6, H6 H; 90-C9; 106-K3; 107-A4, E6 J; 107-J8; 108-A10; 124-B1, E5, H10; 140-K1; 141-A3, C7; 157-F1, H5; 158-A10; 174-A1, C3 NN; 90-A6; WE-F1 W; 106-E1 Y
VA 152 158-C4, J3 H; 157-G5 NN
VA 162 89-F6; WE-A1, B2 W; 90-A7 Y
VA 167 143-B10; 159-B1, B4, B7, C10; 174-K1; 175-A1 H; 174-C4, G3 NN
VA 169 159-F3, J7, K1; 160-B3, B5 H
VA 171 141-B4 NN; 126-F10, K10; 127-A10, E10; 142-J1; 143-E1, J3 Pq; 141-G2; 142-C1 Y
VA 172 142-H5, H8 H; 126-H9; 142-H1 Pq
VA 173 124-D10, G8; 139-K1 NN; 109-F10, K7; 124-K5; 125-A5, E2 Y
VA 199 72-K8; 73-A10; 88-K5; 89-A2, A8; 105-E1; 106-D1 J; 89-D10; 105-E1; 106-D1 W; 73-A6; 90-F10, H8 Y
VA 216 92-K6 G
VA 238 107-J9 NN; 108-A4, F4, K4; 109-B5; YE-A5, D4, G6 Y
VA 273 39-F6, K2 NK
VA 278 143-D10; 159-E1; HE-D2 H
VA 306 157-E2 NN
VA 312 141-C10; 157-C1 NN
VA 321 89-C6, E6, F7; WE-A3, B3 W
VA 322 89-B3 J
VA 337 175-J10 NBN
VA 351 158-G10, K8; 159-A7, H5; 160-A4; 174-F1; HE-B3, G2 H; 174-B2 NN
VA 359 104-E4 J
VA 415 158-F4, K7; 159-A7 H
VA 600 142-B10; 158-B1 H; 56-B5, D1 J; 40-A3, D7, E5, E10 NK; 126-A10; 142-A1, B5 Y
VA 601 39-H9; 54-J5; 55-B5, G2 J; 41-G6, H2 KQ; 40-C7 NK
VA 602 56-K10; 72-K1; 73-A1 J; 73-D4, G7 Y
VA 603 54-J4; 55-A7; 73-A7, D10; 89-D1 J; 38-B5, F10; 54-F1 NK; 89-E4 W; 72-K6; 73-A7, D10; 89-D1 Y
VA 604 73-G4, H7 Y
VA 605 57-C3 J; 41-E2, G3, K3 KQ
VA 606 58-H7, K6 G; 56-D2, H5, K7; 57-A8, H9 J; 41-E1 KQ; 126-C8 Y
VA 607 41-A10; 56-F10, K6; 57-A1, A6; 72-F1 J
VA 608 56-G6 J
VA 610 58-H1, H4 G; 54-K10; 55-G10; 56-A9; 70-J1; 71-D1 J; 103-B10, C10, K9 SC
VA 611 71-G10; 72-C6; 87-G1 J; 126-J6 Pq
VA 612 72-E10, J10; 88-J1; 89-A2 J
VA 613 58-F2, H1 G; 87-G9; 88-A6, D6, J7 J; 125-F9 Y
VA 614 58-K3 G; 72-D10, J6; 88-B5, C1, C10, D9; 104-C1, E3 J; 125-G8, H6 Y
VA 615 70-C3 CC; 159-J3 H; 88-J8; 89-B3, B7; 104-H1, J2 J
VA 616 89-B8 J; 89-B8 W
VA 617 105-D1, D4 J
VA 618 58-K9 G; 105-C3 J; 103-C10 SC
VA 620 55-E6 J; 38-H10, K6; 39-A6 NK; 125-A7, F7, K5; 126-C4 Y
VA 621 70-D6 CC; 39-A10; 55-A1 J; 39-C8 NK; 125-E4, J3; 126-D5 Y
VA 622 54-K2; 55-A2, F2 J; 109-K10; 110-C10; 125-G1 Y
VA 623 54-B9; 70-A9, C1; 86-A2, A7 CC
VA 624 70-A2 CC
VA 626 38-C1 NK
VA 627 70-D5 CC; 38-A2, A4, B5, D10; 54-B6, C1 NK
VA 628 38-A3 NK
VA 629 89-A10; 104-K1; 105-A1 J
VA 630 109-F9; 125-F1, F4 Y
VA 631 56-B10; 71-D3, H4; 72-B1 J; 109-F8, H6 Y
VA 632 72-B6 J; 38-D3, K4; 39-A4, B7, E8 NK; 109-E8 Y
VA 633 92-B1 G; 87-C5, G2; 88-A2 J; 39-E5, F7 NK
VA 634 39-H10; 55-H1 J; 39-A4, C1, H4, J6 NK; 108-K7; 109-A8, D10; 125-C1; YE-D6 Y
VA 635 92-D3 G; 40-C7 NK
VA 636 92-F1 G; 124-H2 Y
VA 637 90-A10; 106-A1 W; 108-B7, F6; YE-A5 Y
VA 639 58-K9 G; 55-J5 J; 39-F3 NK
VA 640 90-G9 Y
VA 641 90-A7; WE-F4 W; 90-A7, E9, J6; WE-F4 Y
VA 644 41-K1 KQ
VA 645 73-D10 J; 41-J4 KQ; 73-D10, J10 Y
VA 646 57-G9; 73-F1 J; 72-K6; 73-A6, D4 Y
VA 647 54-A4 NK
VA 649 56-D10; 72-C2 J; 54-A4, E3 NK
VA 651 126-A4 Y
VA 655 125-J3 Y
VA 656 106-K5 J
VA 657 55-F6 J
VA 658 73-A10; 88-K1; 89-A1 J; 125-J6 Y
VA 659 71-B5 J
VA 662 58-G8 G
VA 665 156-B10 I
VA 666 41-F3 KQ
VA 667 106-K6 J; 41-F4 KQ
VA 669 58-E1 G
VA 671 38-G10; 54-G1 NK
VA 672 38-J10; 54-J1 NK
VA 673 155-A6, A8, D7 I
VA 674 55-H4 J
VA 677 90-D10 J; 90-D10 Y
VA 679 125-J8 Y
VA 680 104-E3 J
VA 681 104-H2 J; 142-A1 Y
VA 682 104-K3; 105-A3 J
VA 684 41-J10; 57-J1 G
VA 690 WE-A1 Y
VA 693 109-C8 Y
VA 696 57-D8 J
VA 699 55-C6 J
VA 704 58-F5 G; 108-J5, K7; 109-A7; YE-D6 Y
VA 705 39-G9 J; 39-G9 NK
VA 706 125-H10; 126-B10; 142-A1 Y

VA 707 155-K10 I
VA 709 125-F5 Y
VA 712 110-F10 Y
VA 713 89-G3 Y
VA 715 54-H10; 55-C10; 70-H1 J
VA 716 90-C8, D4, E5 Y
VA 718 109-E10, H8; 110-A9, B9 Y
VA 746 55-J4; 56-A5 J
VA 755 56-J10; 72-K1 J
VA 759 56-E10 J
VA 766 71-F1 J
VA 780 122-A8 SC
VA 782 126-G8 Pq; 126-E10; 142-E1, G3 Y
VA 786 73-E6 Y
VA 797 73-K4 Y

PLACE NAMES

A

Abbitt Lake 126-F9 Pq
Aberdeen Gdns 158-E6 H
Aberdeen Gdns East 158-F5 H
Aberdeen Gdns North 158-E5 H
Aberdeen Hgts 174-E1 NN
Aberdeen Terr 158-E6 H
Abingdon Ct MHP 92-J3 G
Abingdon Sq 92-J4 G
Acree Acres 125-E3 Y
Adams Hunt 72-F8 J
Addison Ct 158-E7 H
Adrian Ct 159-A7 H
Aldorf Terr 160-A5 H
Allmondsville 58-G7 G
Alpine 141-A8 NN
Andersons Corner 56-A6 J
Archers Mead 106-C3 J
Armistead Hgts 158-H1 H
Armstrong 159-B9 H
Armstrong Pt 159-D8 H
Ashton Green 124-D5 NN
Avery Terr 124-F4 NN
Azalea Gdns 158-K7 H

B

Back River 143-F10 H
Banbury Cross 73-E3 Y
Barclay Woods 140-F9 NN
Barcroft 125-G2 Y
Barhamsville 39-F8 NK
Barlows Corner 73-F3 Y
Baron Woods 88-H9 J
Battle Park 109-B9 Y
Bayberry 141-H5 NN
Beach Road Ests 144-C9 H
Beacon Sq 141-F9 NN
Beacons Bay Tr Pk 92-H3 G
Beaconsdale 157-E1 NN
Beaconsdale Meadows 141-D10 NN
Beechmont 124-C9 NN
Beechwood 124-A9 NN
Beechwood Ests 123-J9 NN
Belleview 41-C3 KQ
Bellgrade 142-E7 H
Bellview Terr 159-G3 H
Bellville 159-A8 H
Bellwood 157-J3 H
Benjamin Clark 73-A8 J
Bennett Farms Ests 126-H7 Pq
Bentley Terr 140-B1 NN
Berkeley Hills 89-C9 W
Berkeleys Green 104-B1 J
Berkley Beach 110-D8 Y
Bernard Vill 141-E8 NN
Bethel Lake 142-B7 H
Bethel Manor 142-A4, C5 Y
Bethel Park 158-D6 H
Betsy Lee Gdns 158-B10 NN
Beverly Hills 140-F10 NN
Bickford Park 124-C10 NN
Bickfords Newton 159-A8 H
Big Bethel 142-C6 H
Big Bethel Hgts 158-B4 H
Birchwood Park 105-C2 J
Blacksmith Corner 107-K3 Y
Blake Terr 158-A3 H
Bonaire 159-K2 H
Bonds Vill 158-E2 H
Boughsprings 105-A1 J
Boulevard 141-E9 NN
Boxley Hills 140-G5 NN
Bradmere 123-H3 NN
Branch Siding 55-C9 J
Brandon Hgts 157-E5 NN
Brandywine 125-J4 Y
Brannock Burn 142-J1 Pq
Breezy Pt 159-F4 H; 125-K8 Y
Brennhaven 140-C5 NN
Brentwood 157-G3 NN
Briarfield Manor 158-C9 NN
Briarfield Vill 158-D7 H
Briarwood Ct 158-G7 H
Briarwood Terr 158-C8 NN
Brighton 141-F5 NN
Brittany Woods 159-K3 H
Broadshore Acres 159-C2 H
Brookhaven 89-A7 J
Brookside 124-B3 NN
Brookside Haven 107-B6 J
Bruton Park 141-G8 NN
Buckroe Beach 160-D4 H
Buckroe Farms 160-A4 H
Buckroe Gdns 160-D2 H
Buckroe Grande 160-C3 H
Buckroe Hgts 160-B4 H
Bunting Manor 126-H10 Pq
Burcher Pt 140-H8 NN
Burkes Corner 73-D4 Y
Burnham Woods 55-J5 J
Burnt Bridge Run 109-B10 Y
Burnt Ordinary 56-B10 J
Burton Woods 72-E10 J
Burts 125-F7 Y
Burwells Glen 106-B3 J
Burwells Green 106-B4 J

C

Calthrop Neck 126-C8 Y
Cambridge Pl 159-G5 H
Camellia Park 126-G9 Pq
Camelot 72-G7 J
Camelot Vill 141-F4 NN
Cannon Park 142-B9 H
Cantamar 160-D4 H
Canterbury Hills 89-A8 J
Capahosic 58-J9 G
Capitol Hgts WE-E2 W
Cardinal Acres 104-F3 J
Carleton Falls 107-H9 NN
Carmines Landing 92-E5 G
Carp Circle 91-J10 Y
Carriage Hgts 89-C5 J
Carriage Hill 140-E1 NN
Carson Hgts 158-A5 H
Carver Ct 159-D4 H
Carver Gdns 90-F10 Y
Carybrook 160-A2 H
Cedar Grove 157-D4 NN
Cedar Hill 123-H1 NN
Cedar Landing Ests 143-A2 Pq
Cedar Park 159-E3 H
Cedar Pt 159-E8; HE-D8 H
Cedarwood 92-J6 G; 124-G8 NN
Celey Park 158-K9 H
Centerville 72-E9 J
Chambrel 89-C5 J
Chanco 89-A10 J
Chancos Grant 104-H2 J
Chapel Park 141-B8 NN
Charles Corner 91-A10; 107-B1 Y
Charleston Hgts 90-D7 Y
Charter Oak 124-F6 NN
Chase Hampton 158-G2 H
Cheadle Hgts 110-A10; 126-A1 Y
Cheatham Annex 91-C4 Y
Chelsea 107-H9 NN
Cherry Acres 159-E5 H
Cherry Creek 124-D3 NN
Chesapeake Vill MHP 140-K3 NN
Chestnut Hill 88-A10 J
Cheyenne Pl 159-D4 H
Chichester Hgts 159-G5 H
Chickahominy Haven 70-G3 J
Chickahominy Shores 54-A7 NK
Chisel Run 73-C10 J
Chiskiake Vill 92-K8 G
Chisman Woods 125-J1 Y
Chismans Pt 110-C10 Y
Christensons Corner 57-G9 J
Christophers Shores 174-F5 NN
Cicero 73-J3 Y
Circle MHP 157-J8 NN
Clairborne Pl 140-K5; 141-A5 NN
Claremont 140-H9 NN
Clark Vill 158-E3 H
Clarkdale 144-B9 H
Claymill Corner 141-B1 NN
Clifton 92-J1 G
Clipper Creek 124-E3 NN
Cliveden 124-F5 NN
Cobble Creek 90-D6 Y
Cobblestone at Lees Mill 123-K3 NN
Coles Forest 123-J9 NN
College Ct 160-B5 H
College Park 140-G3 NN
Colonial Acres 160-C2 H
Colonial Park 90-D9 J
Colonial Terr 90-C10 J; 127-B9 Pq
Colonial Williamsburg WE-D3 W
Colony Meadows 140-B3 NN
Colony Pines 124-H4 NN
Compton Pl 144-C7 H
Concord Lake 141-H6 NN
Conover 58-K9 G
Conway Garden 89-C10 J
Cornwall Terr 157-J4 H
Counselors Close 89-G8 W
Country Club 143-C10; 159-B1 H
Country Club Acres 106-G1 Y
Country Vill MHP 107-C6 J
Courthouse Green 124-D8 NN
Courtney Trace 141-H10 NN
Cove Homes 125-J8 Y
Crandols Landing 142-K1 Pq
Crestwood 159-F4 H
Criston 141-A3 NN
Croaker 56-J7 J
Crown Grant 140-D2 NN
Crystal Lake 158-J6 H
Curle Neck 159-G3 H
Cypress Crest 92-F4 G
Cypress Pt 54-H10 J

D

Dandy 110-A7 Y
Dare 126-A3 Y
Davis MHP 157-J8 NN
Daybreak 124-H9 NN
Days Pt 155-C6 I
Deep Creek 140-H7 NN
Deer Park 141-C9 NN
Deer Park Grove 141-D8 NN
Deerfield 141-F5 NN
Deerwood Hills 87-H1 J
Denbigh 124-D9 NN
Denbigh Bluffs 124-B8 NN
Denbigh Plantation 140-D7 NN
Denbigh Shores 140-B1 NN
Denbigh Terr 140-B1 NN
Denbigh Trace 124-C4 NN
Denbigh Vill 124-G10 NN
Denbrook Sta 124-K7 NN
Derby Run 142-G6 H
Devon Pl 140-K7 NN
Devonshire 125-K7 Y
Diascund 55-A7 J
Douglas Park 158-C7 H
Druid Hills 89-A10 J
Drummonds Corner 142-G6 H
Drummonds Field 104-D3 J
Dunbar Gdns 158-J9 H
Dunlap Woods 92-G5 G
Dunmore 125-C2 Y
Durfeys Mill 105-B2 J
Dutch Vill 157-E3 NN

E

Eagle Sound 141-F1 Y
East End 174-G2 NN
East Hampton 159-H6; HE-H2 H
Easthill Ests 160-C2 H
Eastwood 140-F1 NN
Edgarton 160-A3 H
Edgehill 109-B10 Y
Edgemoor 107-H9 NN
Edgewater 144-B8 H
Edgewood 140-H9 NN; 141-J1 Y
Edinburgh Farms 143-K8 H
Edloe Terr 107-K5 Y
Edrale Pl 160-A5 H
Elizabeth Lake Ests 159-H3 H
Elliott Pl 144-B7 H
Elmwood 56-G7 J
Endview 107-J6 NN
Endview Woods 108-A4 Y
Essex Park 160-B1 H
Evergreen Shores 110-D10 Y
Ewell 73-A8 J
Ewell Hall 73-A10 J

F

Fairfax Woods 105-K4 J
Fairfield 160-A1 H
Fairfield Williamsburg 89-H5 Y
Fairview 156-K2 NN
Fairview Farms 159-C3 H
Fairway 140-K10 NN
Fairway Vista 88-D9 J
Farmington 157-J3 H
Fenton Ests 57-D8 J
Ferguson Glade 141-J4 Y
Fieldcrest 88-C10 J
First Colony 104-B2 J
First Settlers Landing 104-C3 J
Five Forks 88-H10 J
Floyds Bay 127-A8 Pq
Fordham 160-A4 H
Fords Colony 88-E2 J
Forest Glen 72-E10 J
Forest Grove 107-C6 J
Forest Hill Park 89-K5; WE-F1 W
Forest Park 158-B5 H
Forrest Park 142-H2 Pq
Forrest Ridge 158-A3 H
Fort Magruder Hgts 90-B8 J
Four Mile Tree 103-D8 SC
Four Seasons 142-E5 Y
Fox Corner 158-B4 H
Fox Hill 144-C8 H
Fox Hill Shores 144-F8 H
Fox Ridge 72-E8 J
Foxbourne Pl 144-A10 H
Foxbridge 143-K10 H
Foxfield 159-K1 H
Foxhaven 58-J9 G
Freemans Gdns 127-C10 Pq
Freemoor Ests 126-E8 Pq

G

Gainesville 158-H10 H
Garden Vill 159-H4 H
Gardners MHP 142-K9 H
Garrett Ct 158-D4 H
Gate House Farms 104-K4 J
Gateway 159-B1 H
Georgetown Commons 124-C8 NN
Georgian Manor 126-G9 Pq
Glen Garden 158-B7 NN
Glen Laurel 125-E5 Y
Glendale 141-B9 NN
Gleneagles 125-B10 NN
Glenwood Acres 56-J7 J
Gloucester Pt 108-J1 G
Goffigan Gdns 109-C9 Y
Goodwin Neck Ests 110-B6 Y
Governors Sq 89-B7 W
Grafton 125-G4 Y
Grafton Branch 125-G4 Y
Grafton Sta 125-G5 Y
Grafton Woods 125-E5 Y
Grandview 144-E8 H
Grandview Shores 144-E8 H
Graves Ordinary 106-D5 J
Graylin Woods 88-J10 J
Great Woods 56-D6 J
Green 160-B1 H
Green Acres 124-G4 NN
Green Meadow 124-C10 NN
Green Oaks 157-F3 NN
Green Swamp 88-C4 J
Greenbriar 158-G10 H
Greenfield Vill 159-A2 H
Greensprings 89-G4 Y
Greensprings MHP 88-A6 J
Greensprings Plantation 88-C6 J
Greenway Farms 143-F10 H
Gressitt 41-G3 KQ
Greyhound Ests 73-A10 J
Griffins Beach 126-J6 Pq
Grove Hill Ests 55-D9 J

H

Halloway Bluffs 124-A9 NN
Hallwood 144-C10 H
Hampton 142-E9; 143-H8; 158-E3; 159-C4; 174-J2; 175-A1 H
Hampton Club 158-F2 H
Hampton Commons 158-G2, H6 H
Hampton Harbour 159-G6; HE-F5, G5 H
Hampton Roads 159-B10 H
Hampton Roads Ctr 142-C10; 158-E1 H
Hampton Shores 159-H3 H
Hampton Terr 158-J7 H
Hampton Woods 158-B1 H
Hampton Woods Condo 158-C1 H
Hanover Hgts 124-G8 NN
Hansom Park 126-H7 Pq
Happy Island 144-E8 H
Harbor Hills 92-J7 G
Harbor Terr 110-A8 Y
Harborview Ests 140-G6 NN
Harper Woods 141-J6 NN
Harpersville 141-H6 NN
Harpersville Hgts 141-H10 NN
Harris Grove 109-D9 Y
Harwood 107-A5 J
Harwood Hgts 125-H8 Y
Harwood Mill 125-H8 Y
Harwood Mill Tr Ct 125-G7 Y
Hayes 92-K7 G
Heartwood 160-B2 H
Heather Lakes 159-A1 H
Heiland Gdns 141-G9 NN
Heritage Cove 126-E8 Pq
Heritage Hamlet 125-H1 Y
Heritage Landing 88-A10 J
Heritage Tr Pk 107-A6 J
Hickory Hill 140-K7 NN
Hickory Hills 108-F3; YE-A3 Y
Hickory Pt 124-A6 NN
Hidden Ests 140-J7 NN
Hidenwood 141-B10 NN
Highland Park 89-H5; WE-C1 W
Hilton Pl 157-J7 NN
Hilton Vill 157-G6 NN
Holdcroft 70-A2 CC
Holiday Park 143-K9 H
Hollingsworth 125-D10 Y
Holloway Forest 143-A1 Pq
Holloway Forrest 127-A10 Pq
Holly Bank 140-G7 NN
Holly Forks 40-E6 NK
Holly Hills 89-E10 W; 125-G2 Y
Holly Hills Carriage 89-D10 W
Holly Homes 159-K3 H
Holly Mead 141-K1 Y
Hollybrook 105-A2 J
Hollywood Ests 125-J8 Y
Hoopes Landing 139-K1 NN
Horizon Plaza 159-B6 H
Hornsbyville 109-D9 Y
Horse Br 86-A6 CC
Horse Run 140-J2 NN
Horse Run Creek 140-G1 NN
Horsepoint Farms 140-E5 NN
Howards Landing 126-D5 Y
Howe Farms 143-J8 H
Hudson Terr 157-H4 NN
Hultwood 140-J7 NN
Hunt Woods 142-G2 Pq
Hunters Creek 72-B2 J
Hunters Glenn 140-K5 NN
Huntington Hgts 157-H9 NN

I

Indian River Park 159-A10 H
Indian Springs 89-F8; WE-B4 W
Indigo Park 89-A9 J
Indigo Terr 89-A7 J
Inlet Pt 144-B10 H
Institute Gdns 159-J5 H
Ironbound Sq 89-B5 J
Ivy Dell 57-C2 J
Ivy Farms 157-G1 NN
Ivy Home 159-E8 H

J

Jacobs Springs 125-H5 Y
James Landing 156-H2 NN
James River Hgts 155-K10 I
James Shire Settlement 72-G6 J
James Sq 105-C1 J
James Terr 90-C8 J
James York MHP 90-B9 J
Jamestown 104-F7 J
Jamestown 1607 104-F2 J
Jamestown Farms 88-K9 J
Jamestown Manor 157-D3 NN
Je Mar Hgts 140-G8 NN
Jefferson East 141-E8 NN
Jeffersons Hundred 106-G5 J
Justinian Grove 141-J3 Y

K

Keller 58-H7 G
Kentucky Farms 125-A7 Y
Key Oak 126-J10 Pq
King Sq 159-E4 H
King Street Commons 159-E4 H
Kings Corner 55-A6 J
Kings Creek Plantation 90-J8 Y
Kings Ct 108-G4; YE-A5 Y
Kings Park 140-D4 NN
Kings Pt 159-G4 H
Kings Vill 39-H10 J
Kings Villa 125-H10 Y
Kingslee Park 159-E1 H
Kingsmill 106-C2 J
Kingspoint 105-H3 J
Kingstowne 141-C7 NN
Kingswood 105-A1 J
Klondike 159-K7 H
Knollwood Meadows 123-H10 NN
Kristiansand 72-F2 J

L

La Tierra 126-H9 Pq
Lackey 108-B4 Y
Lafayette Ests 92-J10 G
Lafayette Sq 88-J1 J
Lake Hampton 159-A5 H
Lake Park 158-H10 H
Lake Toano Ests 72-A1 J
Lakehaven 158-E3 H
Lakes at Dare 126-B3 Y
Lakeside 158-F4 H; 141-C2; 157-B2 NN
Lakeside Forest 125-H6 Y
Lakeside Hgts 125-J7 Y
Lakeside Homes 125-H7 Y
Lakeview 125-K5 Y
Lakewood 104-K2 J
Lakewood Park 140-B4 NN
Lambs Creek 126-D9 Y
Landsdowne 140-C3 NN
Lanexa 54-E3 NK
Langley Circle 159-C4 H
Langley Ct 159-D1 H
Langley Hgts 159-D3 H
Langley MHP 142-J7 H
Langley Park 159-D5; HE-A1 H
Langley View 143-G8 H; 143-A9 L
Laser Vill 158-A6 H
Lassiter Ct 174-E5 NN
Lawndale Farms 140-F6 NN
Lawson 89-E5 W
Lee Hall 107-H10 NN
Lee Hgts 158-F6 H
Lees Mill 123-K4 NN
Lees Vill 125-H8 Y
Lewishaven 158-F8 H
Lexington 125-C10 Y
Liberty Terr 159-J1 H
Lightfoot 72-J6 J; 73-A6 Y
Lincoln Park 159-C6; HE-A4 H
Linda Ct 92-H1 G
Little Creek 160-B1 H
Little England 159-B7 H
Little Farms 158-K10 H
Littletown Quarter 106-D4 J
Lochaven 160-C3 H
Lochaven Pl 160-B4 H
Locust Hill 58-K4 G
Locust Run 127-C10 Pq
Lombardy 124-B10 NN
Longhill Gate 88-H1 J
Longhill Sta 72-E8 J
Longhill Woods 89-D3 W
Lotz Acres Ests 142-B1 Y
Lowneyville 58-J2 G
Lucas Creek Park 140-C1 NN
Lyliston 157-H2 NN
Lynnhaven 158-D2 H
Lynnhaven Shores 158-E3 H

M

Macalva Park 159-E2 H
Madison Chase 142-F7 H
Magruder Commons 158-F2 H
Magruder Ests 142-F6 H
Magruder Hgts 158-G1 H
Malo Beach 160-E2 H
Malvern 159-F2 H
Mammoth Oaks 140-H7 NN
Manoroe Colony 159-J7 H
Maple Grove 140-G4 NN
Marina MHP 126-K8 Pq
Marina Pt 123-J10 NN
Mariners Pt 158-C5 H
Marlbank 109-D7 Y
Marlbank Cove 109-D9 Y
Marlbank Farm 109-C7 Y
Marstons MHP 72-G6 J
Masons Grant 89-A7 J
Maury Pl 157-D2 NN
Maxwell Gdns 140-J6 NN
Maymont 141-A6 NN
Meadow Brook 159-C9 H
Meadow Lake 55-H5 J
Meadowlake Farms 141-K1 Y
Meadowland 141-G10 NN
Meadowview Abraham 158-B7 NN
Menchville 140-E5 NN
Menchville Meadows 140-G5 NN
Menzels 71-B4 J
Mercury Central 158-H1 H
Mercury West 158-C5 H
Meredith Woods 140-H5 NN
Merrimac Shores 159-E9 H
Merry Oaks 55-G7 J; 124-A4 NN
Merry Pt Ests 156-H2 NN
Merry View 156-J1 NN
Messick Terr 143-F1 Pq
Mews at Williamsburg 89-B2 J
Michaels Woods 157-J1 H
Michaels Woods of Northampton 158-A1 H
Michelle Landing 126-F7 Pq
Middle Plantation 88-F1 J
Middle Towne Farms 90-B7 Y
Mill Creek Landing 160-D5 H; 89-A8 J
Mill Creek Terr 159-J8 H
Mill Farms 125-H9 Y
Mill Pt 159-F6; HE-E4, F4 H
Millers Cove 140-B4 NN
Millside 125-H3 Y
Mirror Lake Ests 56-F9 J
Mobile Ests 72-J5 J
Moodys Run 106-C4 J
Moores 126-E10 Y
Morgarts Beach 155-E5 I
Morning View 160-D3 H
Morris Manor 158-C7 H
Morrison 141-F10 NN
Moss Green 105-G1 J
Mount Airy 70-C4 CC
Mount Folly 57-B1 J
Mount Pleasant 103-H8 SC
Mount Zion 70-B8 CC
Mulberry Hill 139-J1 NN

N

Nelson Farms 142-C10 H
Nelson Hgts 109-B8 Y
Nelson Park 90-C8 Y
Nelson Pl 157-D4 NN
Neva Terr 92-H5 G
New Bridge 142-G3 Pq
New Hampton Vill 159-A8 H
Newgate Vill 142-B6 H
Newmarket Vill 158-A7 NN
Newport 124-G9 NN
Newport News 108-A8; 122-J10; 124-C7; 125-A10; 140-F6; 141-C6; 156-F2; 157-D4; 173-H1; 174-C2 NN
Newport Sq 141-G7 NN
Newsome Park 174-C1 NN
Nicewood 140-D3 NN
Norfolk 175-F9 N
Norge 72-H2 J
Norge Ct 72-F2 J
Normandy Lake 140-J7 NN
North Cove 73-E1 J
North Hilton 157-F4 NN
North Newport News 157-H7 NN
Northampton 158-C2 H
Northampton Vill 157-J5 H
Nottingham Vill 140-E4 NN
Nubian Park 159-J3 H
Numerous 140-A1 NN

O

Oak Park 157-F2 NN
Oak Tree Farms 58-K1 G
Oakdale 92-J6 G
Oakland 159-H6 H; 72-E2 J
Oakmoore 126-F8 Pq
Oaktree 73-G7 Y
Old Farm Ests 141-C8 NN
Old Pt Vill 159-J5 H
Old Quaker Ests 73-C2 Y
Old Stage Manor 73-D8 Y
Old Towne 159-E6; HE-C4, D4 H
Oxford 159-F5 H
Oyster Pt 140-K3 NN

P

Padgetts Ordinary 106-D4 J
Page Landing 104-K5 J
Page Pl 126-J10 Pq
Paradise Pt Ests 126-K9 Pq
Park Pl 158-G9 H; 140-B3 NN
Parkview 158-A9 NN
Parkway 90-A7 W
Parkway Ests 90-C6 Y
Pathco MHP 157-K9 NN
Patricia Hgts 124-A5 NN
Patrician Manor 158-C3 H
Patricks Landing 126-B6 Y
Patriot Vill 142-A4 Y
Patriots Colony 88-A10 J
Patriots Sq 109-C10 Y
Pauls Tr Pk 124-C8 NN
Pavilion Ests 158-G6 H
Pear Tree Hall 140-F1 NN
Pelegs Pt 104-J5 J
Penniman East 90-F9 Y
Peppertree 89-B9 J
Phenix Terr 159-C4 H
Phillips Lake 159-G3 H
Phoebus 159-J7 H
Pile 155-A5 I
Pine Chapel Vill 158-K5 H
Pine Grove Vill 141-G8 NN
Pine Hill 58-J1 G
Pine Tree Pl 158-A4 H
Pinecrest WE-F2 W
Pinecroft 141-A8 NN
Pinegrove Ct 159-F1 H
Pinehurst 158-J9 H
Pineridge 72-J1 J
Pines of York 142-C4 Y
Pinetta 58-J4 G
Pinewood 126-H6 Pq
Piney Creek Ests 89-B2 W
Plantation Acres 125-J10 Y
Plantation Hgts 89-G4 Y
Pocahantas 124-F9 NN
Pollard Park WE-B4 W
Pools Grant 159-F6; HE-F3 H
Poplar Creek 72-H1 J
Poplar Hall Plantation 107-C6 J

Poquoson 126-E10; 127-D9; 142-H2; 143-F2 Pq
Poquoson Pl 126-G9 Pq
Poquoson River Ests 126-E6 Pq
Poquoson Shores 126-G5 Pq
Port Anne 89-F9 W
Powhatan 92-G4 G
Powhatan Chimney 92-E4 G
Powhatan Crossing 88-F9 J
Powhatan of Williamsburg 88-F7 J
Powhatan Park 158-G8 H; WE-F4 W
Powhatan Pl 126-F9 Pq
Powhatan Plantation 88-F9 J
Powhatan Shores 104-H4 J
Presidential Park 141-H8 NN
Prestige Park 158-G6 H
Priorslee 90-B6 W

Q

Quail Hollow 92-J2 G; 125-F6 Y
Quartermarsh Ests 125-J6 Y
Queens Lake 90-D5 Y
Queens Terr 159-B6 H
Queenswood 90-B7 Y
Quiet Vill 141-F8 NN

R

R & L MHP 92-H2 G
Racefield 55-E2 J
Raleigh Sq 104-J3 J
Raleigh Terr 159-B10 H
Randolphs Green 106-C4 J
Red Hill Vill 124-H10 NN
Regents Walk 158-C2 H
Rich Acres 141-G1 Y
Rich Neck Hgts 89-C9 W
Richfield 124-F5 NN
Richmond Park 160-D3 H
Richneck Ests 124-G6 NN
Ridgecrest 157-K3 H
River Land 159-E2 H
River Mews 123-K6 NN
River Pt 143-H9 H
River Terr 143-E10 H
River View Plantation 57-K9 J
Rivercreek Acres 126-H6 Pq
Riverdale 159-A3 H
Rivergate 126-J8 Pq
Riverhaven 126-B9 Y
Riverpoint West 143-H9 H
Riverside 157-A2 NN
Riverview Park 58-G6 G
Roane 41-G9 KQ
Robanna Shores 125-K2 Y
Robert Acres 126-F7 Pq
Roberta Terr 158-C4 H
Roberts Landing 126-G6 Pq
Roberts Run 126-G7 Pq
Roberts Trace 157-J4 H; 142-E1 Y
Robinson Terr 141-H9 NN
Rochambeau Vill 107-H6 Y
Rock Creek 141-D1 Y
Rodgers Villa 160-C3 H
Rolling Woods 105-B2 J
Rosalee Gdns 158-K8 H
Rosevale 160-A6 H
Rosewood 92-J2 G
Royal Grant 90-B5 Y
Running Man 142-D2 Y
Runnymede 140-D4 NN
Russell Terr 142-B8 H
Rustic 86-A3 CC

S

Saddle Creek 140-D5 NN
Saddletown 57-C9 J
Sadie Lee Taylor 106-K6 J
Saint Andrews 124-F8 NN
Saint Georges Hundred 104-E1 J
Saint James Pl 124-J7 NN
Salt Pond 160-D2 H
Sandhill 56-D7 J
Sandy Hill Tr Ct 92-F3 G
Sanlun Lake 142-B7 H
Sassafras Landing 58-H5 G
Sawmill Corner 107-C2 Y
Schenck Terr 73-G8 Y
Scotch Toms Wood 125-H3 Y
Scotland Sq 159-K2 H
Seaford 110-A9 Y
Seaford Shores 110-C9 Y
Seasons Trace 72-J10 J
Sedgefield 157-H4 NN
Seldendale 159-B3 H
Settlers Crossing 109-B9 Y
Settlers Mill 104-J2 J
Seven Hollys 126-C3 Y
Shady Banks 142-C3 Y
Shady Oaks MHP 142-H3 Pq
Sharon 140-H1 NN
Shellbank Woods 104-A1 J
Shellis Sq 90-A6; WE-F1 W
Sherwood Park 159-D1 H
Shilsons Corner 92-B10 Y
Ship Pt 126-E3 Y
Shore Park 140-B2 NN
Sierra Park 141-F8 NN
Silver Isles 144-D10 H
Sinclair Circle 158-K3 H
Sinclair Farms 159-D2 H
Sinclair Manor 142-C1 Y
Sixty South 124-B7 NN
Skiffes Creek Annex 107-F6 NN
Skiffes Creek Manor 107-C7 J
Skiffes Creek Terr 107-C7 J
Skillman Est 55-F4 J
Skimino 73-H5 Y
Skimino Farms 73-K4 Y
Skimino Hills 73-F4 Y
Skimino Landing 73-J2 Y
Skipwith Farms 89-C2 W
Slimedes 140-H9 NN
Smithville Terr 141-H1 Y
Sommerville 109-K10 Y
Sonoma Woods 124-F7 NN
Sonshine Acres 108-A5 Y
South Lee Hall 107-J10 NN
South Rivermont 157-G7 NN
Southall Acres 160-D1 H
Southall Landing 160-E1 H
Southampton 159-C8 H
Southlake 141-B2 NN
Spinnaker Cove 142-H6 H
Spinnaker Shores 158-E4 H
Spring House 124-G10 NN
Spring Trace 141-J6 NN
Springfield Terr 90-E7 Y
Springhill 88-E5 J
Station Ests 157-K5 H
Steeple Chase 88-H8 J
Stonehurst Terr 159-F4 H
Stoneybrook 124-A7 NN
Stratford Hall 89-C1 J
Stratford Terr 157-H5 NN
Stuart Gdns 174-F4 NN
Sulik MHP 142-G5 H
Summerlake 140-H5 NN
Sussex Hilton 158-B9 NN
Swansea Manor 141-J7 NN
Swantown 157-G4 NN
Sweetbriar 141-B10 NN
Sycamore Landing 41-A10 J
Sylvia Est 158-D6 H
Sylvia Ests 158-B5 H

T

Tabb 125-G10; 141-G1 Y
Tabb Lakes 141-J3 Y
Tabb Terr 141-J1 Y
Tall Pines 140-K8 NN
Tanglewood 158-J3 H
Tazewells Hundred 106-B3 J
Temple Hall Ests 56-C5 J
The Arbors 124-D6 NN
The Cascades 141-A1 NN
The Colonies 54-F7 NK
The Colony 105-A3 J
The Coves 89-F9 W
The Fairways 141-B2 NN
The Foxes 88-J8 J
The Governors Land at Two Rivers 103-H1 J
The Hamlet 89-B1 J
The Harbours 141-C7 NN
The Landing 106-A5 J
The Links 88-E3 J
The Meadows 88-J8 J
The Mews 89-B2 J
The Midlands 89-B8 W
The Oaks on Henry 89-G10 W
The Twp in Hampton Woods 158-G3 H
The Villages at Woodside 124-E5 NN
The Vineyards at Jockeys Neck 105-E4 J
The Woods 89-D10 W
The Woods of Tabb 142-F2 Y
The Wyndham 158-G2 H
Thomas Vill 158-H5 H
Thornhill 158-F7 H
Threechopt Vill 158-C5 H
Tide Mill Ests 126-A10 Y
Tidemill Commons 142-H10 H
Tidemill Farms 142-J9 H
Timberneck Ests 92-F1 G
Toano 56-A9 J
Toano Terr 56-A10 J
Toano Trace 72-B1 J
Todd Pt 140-G10 NN
Todds Colony 158-A4 H
Todds Lane Manor 157-J4 H
Topping Landing 142-K2 Pq
Torrey Pine 124-B6 NN
Towering Pines 126-G9 Pq
Town Villas 126-G9 Pq
Town Villas South 126-G10 Pq
Towne Sq 157-H8 NN
Tradewinds 142-F7 H; 141-E1 Y
Trailux Mobile Vill 124-E8 NN
Treasure Pt 159-F4 H
Tuckahoe 140-K7 NN
Tucker Farm 142-J1 Pq
Turnberry Wells 124-G9 NN
Turner Terr 157-J4 H
Tutters Neck 106-A2 J

V

Valmoore 126-G8 Pq
Valmoore Est 126-F8 Pq
Vandale 158-A6 H
Victoria Sq 159-C7 H
Victoria Sta 124-K6 NN
Village Green 141-C3 NN; 89-D10 W
Village Landing 92-F2 G
Village of Williamsburg 90-A9 W
Village Sq 104-J1 J
Villages at Wesminster 72-K8; 73-A8 J
Villages of Kiln Cr 141-C1 NN
Virginia Hgts 158-E8 H

W

Walden 140-G2 NN
Wales 89-D4 W
Walker Pl 157-J5 NN
Walkers 54-B4 NK
Walkers Landing 143-E10 H
Walnut Hills 89-D8 W
Walnut Homes 159-K3 H
Ware Creek Manor 57-A4 J
Warren Mill 71-F1 J
Warwick 157-F4 NN
Warwick Gdns 158-B10 NN
Warwick Lawns 124-G10 NN
Warwick Mobile Home Ests 141-A5 NN
Warwick on the James 157-D5 NN
Warwick River Ests 123-J10 NN
Warwick Shores 140-D6 NN
Warwick Villa 141-F10 NN
Warwickshire Ct 157-H5 NN
Waterford 88-G8 J
Watergate Ests 140-F10 NN
Waters Ridge 107-H10 NN
Weade Terr 142-B9 H
Wellington Pl 157-K4 H
Wendwood 140-J4 NN
West End 41-J10 G
West Hampton 159-C7 H
West Over Shores 126-E9 Pq
West Williamsburg WE-A1 W
Westfield Vill 158-G10 H
Westgate 141-A1 NN; 89-C2 W
Westmoreland 88-K1 J
Westover 159-C2 H
Westover Pl 159-A9 H
Westray Downs 88-K10 J
Westview Green 141-J8 H
Westview Lakes 142-A7 H
Westwood 158-A3 H
Wexford Hills 57-E9 J
Whealton Terr 157-K5 H
Wheatland Points 159-A3 H
Whispering Pines 126-A6 Y
Whispering Winds 142-E5 Y
White Acres 158-A4 H
White Oak 157-D3 NN
White Oaks 88-K9 J
Whitehouse Cove 127-A8 Pq
Whittaker Mill 106-D2 J
Wickhams Grant 106-E4 J
Wicomico 92-J4 G
Wilken Park 158-K5 H
Williamsburg 89-C7; 106-A1; WE-E4 W
Williamsburg Bluffs 90-F10 Y
Williamsburg Commons 89-F5; WE-B2 Y
Williamsburg Landing 105-D1 J
Williamsburg Terr MHP 90-D9 J
Williamsburg West 88-J2 J
Willow Beach 159-K8 H
Willow Green 141-F4 NN
Willow Oaks 159-H1 H
Willow Pt 141-C1 NN
Willowood 140-D2 NN
Wimbleton Terr 158-C3 H
Winchester Park 158-D3 H
Windbrook 141-A2 NN
Windemere Farms 140-B4 NN
Winders Pond 125-J5 Y
Windsor Forest 88-H2 J
Windsor Great Park 124-H7 NN
Windsor Terr 159-A4 H
Windward Towers 157-H8 NN
Windy Hill MHP 107-B6 J
Windy Pt 142-K1 Pq
Wingfield 124-B8 NN
Winster Fax 106-D3 J
Winston Terr 89-B10 J
Winterhaven 141-B7 NN
Wolftrap Ests 125-G3 Y
Wood Towne Quarters 125-B1 Y
Woodcreek 124-C3 NN
Woodhaven 124-C6 NN
Woodlake Crossing 141-J2 Y
Woodland Farms 142-B8 H; 57-C5 J
Woodland MHP 125-F3 Y
Woodland Park 159-J4 H
Woodmere 140-H10 NN
Woods of Williamsburg 72-K10 J
Woods Run 124-E3 NN
Woodscape 124-D5 NN
Woodside Terr 124-F3 NN
Wyndham Plantation 90-A5; WE-F1 W
Wynns Grove 107-A5 J
Wythe 158-H10 H
Wythe Creek 126-H10 Pq
Wythe Creek Farms 142-E2 Y
Wythe Farms 158-D4 H
Wythe Pl 174-K1 H
Wythe Terr 158-J10 H

Y

Yachthaven Terr 160-C5 H
York Crossing 125-H6 Y
York Haven Anchorage 126-J8 Pq
York Manor 141-G2 Y
York Meadows 141-H4 Y
York Plaza 140-E1 NN
York Pt 126-G2 Y
York River 90-K1 Y
York River Pines 92-J9 G
York Terr 90-E8 Y
York View 92-K6 G
Yorkshire 89-E10 W; 125-H10 Y
Yorkshire Downs 142-E4 Y
Yorkshire Park 110-A10 Y
Yorktown 108-J2; YE-E2 Y
Youngs Mill 140-H3 NN

Z

Zooks Tr Ct 108-E3 Y

AIRPORTS

Camp Peary Landing Strip 90-H1 Y
Felker Army Airfield 123-C9 F
Fort Monroe Walker Airfield 160-A8 FM
Newport News/Williamsburg Intl 124-J10; 125-A9; 140-J1; 141-A1 NN; 125-A9 Y
Newport News/Williamsburg Intl Airport Terminal 124-K10 NN
Williamsburg-Jamestown 105-E2 J

BRIDGES

B T Washington Mem Br HE-E5 H
Chickahominy Br 87-A8 J
Chickahominy Bridge 86-K8 CC; 86-K8 J
Diascund Br 54-C2 NK
George P Coleman Mem Br 108-J2 G; 108-J2; YE-D1 Y
Hampton Roads Br-Tunnel 159-J9; 175-K1 H
James River Br 173-B3 I; 157-E10; 173-B3 NN
James V Bickford Mem Br HE-F2 H
Jamestown Ferry 104-D7 J; 104-D7 SC
Langley Br 143-D9 H; 143-D9 L
Lee Hall Res Dam #1 124-D1 NN
Little Cr Dam 71-F4 J
Monitor-Merrimac Mem Br-Tunnel 174-E7 NN
Swamp Br 108-F10 Y

BUSINESS PARKS

American Oil Refinery 109-H7 Y
Ashe Ind Pk 141-H1 Y
Bethel Ind Pk 141-G3 Y
Busch Commerce Pk 90-G8 Y
Busch Corp Ctr 106-E2 J
Busch Ind Pk 90-F8 Y
Camp Morrison Ind Pk 157-F3 NN
Colony Square of Denbigh 124-H8 NN
Copeland Ind Pk 158-D9 H; 174-E1 NN
Dare Prof Pk 125-G4 Y
Ewell Ind Pk 73-C9 Y
Governor Berkeley Prof Ctr 89-B10 W
Green Mount Ind Pk 107-D8 J
Greene Ind Pk 125-C1 Y
Hampton Ind Ctr 159-A7 H
Hampton Ind Pk 158-G8 H
Hampton Roads Ctr 142-C10; 158-E1 H
Hankins Ind Pk 56-C9 J
Ivans Ind Pk 141-B6 NN
James River Commerce Ctr 107-B8 J
John Tyler Commercial Pk 88-J9 J
Kiln Cr Corp Ctr 141-E2 Y
Kiln Cr Ctr 141-G4 Y
Langley Research & Development Pk 142-G7 H
McCale Prof Pk 141-F7 NN
New Quarter Ind Pk 88-K6; 89-A6 J
Newport News Ind Pk 158-D10; 174-C1, D1 NN
Newport News Marine Terminal 174-B4 NN
Newport News Ship Bldg & Drydock Company 157-H10; 173-J1 NN
Oakland Ind Pk 123-G2 NN
Oyster Pt of Newport News 141-D6 NN
Patrick Henry Bus Ctr 124-J8 NN
Patrick Henry Commerce Ctr 124-J8; 125-A8 NN
Peninsula Ind Pk 124-E4 NN
Seafood Ind Pk 174-D6 NN
Stonehouse Commerce Pk 56-A4 J
Victory Ind Pk 125-D2 Y
Williamsburg Office Pk 89-A10 J
Wythe Cr Ind Pk 142-G5 H
York River Commerce Pk 109-D10; 125-D1 Y

CAMPSITES

Anvil 89-E2 Y
Camp Skimino 73-C1 Y
Colonial 73-G3 Y
Fair Oaks at the Pottery 73-A4 Y
First Settlers 104-G4 J
Five Forks 88-G10 J
Jamestown Beach 104-D4 J
Kin-Kaid 73-D5 Y
Newport News Pk 108-B10; 124-B1 NN
Norfolk Camp 39-B9 NK
Outdoor World 56-G8 J
Peninsula Boy Scout Resv 72-D6 J
Powhatan Resorts 87-A8 J
Williamsburg 72-G4 J
Williamsburg KOA 73-G2 Y

CEMETERIES

Barnes 159-H5 H
Basset 159-E5; HE-C3 H
Cedar Grove 89-G9 W
Cheesecake 107-A1 Y
Chickahominy Bapt 71-J4 J
Clark 144-B10 H
Curtis 107-B4 J
Eastern 143-G1 Pq
Eastern St 89-G9 W
Ebenezer 54-C3 NK
Elmerton 159-E5; HE-D2 H
Emmaus Bapt 126-G8 Pq
French 108-F7 Y
Gloucester Field 92-F2 G
Good Hope 38-A4 NK
Greenlawn Mem Pk 174-F1 H; 174-F1 NN
Hampton Mem Pk 142-G10; 158-G1 H
Jackson Post 107-G4 Y
James River Bapt 88-B2 J
Jamestown Presb 88-G10 J
Lee 107-K2 Y
Liberty Bapt 54-G2 NK
Magruder 74-B9 Y
Morning Star Bapt 107-D7 J
Mount Nebo 39-E3 NK
Mount Olive 40-D4 NK
Mount Pilgrims 73-G5 Y
Mount Pleasant Bapt 70-A2 CC
Mount Zion Bapt 70-A8 CC
National 159-G7, H7; HE-G7, H6 H; 109-C5 Y
New Zion Bapt 72-F10 J
Oakland 159-H5; HE-H2 H
Our Savior 72-F1 J
Parklawn Mem 158-J2 H
Peninsula Mem Pk 140-K6; 141-A5 NN
Rosenbaum Mem Pk 158-J10 H
Saint Johns 159-E6; HE-C4 H
Shiloh 56-K7 J
Smith 142-J2 Pq
Smith Mem Lightfoot 72-J6 J
Swanns Pt Plantation 103-K8 SC
Tabernacle 40-B7 NK
Tabernacle United Meth 126-K9 Pq
Travis 104-F7 J
Weston 143-E3 Pq
Williamsburg Mem Pk 73-B10 J
Williamsburg Mennonite 56-H10 J
Yorktown Natl 108-K4; YE-E5 Y
Zion 72-K7 J
Zion Prospect 125-F9 Y

COLLEGES & UNIVERSITIES

Christopher Newport Univ 140-K10; 141-A10; 157-B1 NN
Coll of William & Mary 92-J10; 108-J1; 109-A1 G; 89-D7; WE-A3, C4 W
Coll of William & Mary Alumni House WE-B3 W
Coll of William & Mary Blow Mem Hall (Admin) WE-B3 W
Coll of William & Mary Botetourt Complex WE-A3 W
Coll of William & Mary Brafferton WE-C4 W
Coll of William & Mary Bryan Complex WE-B3 W
Coll of William & Mary Campus Ctr WE-C4 W
Coll of William & Mary Cary Field Zable Stadium 89-F7 W
Coll of William & Mary Coll Bookstore WE-C4 W
Coll of William & Mary Commons WE-A3 W
Coll of William & Mary Dupont Hall WE-A4 W
Coll of William & Mary Ewell Hall WE-B3 W
Coll of William & Mary Fraternity Complex WE-A3 W
Coll of William & Mary Hall WE-A3 W
Coll of William & Mary James Blair Hall WE-B3 W
Coll of William & Mary Jones Hall WE-A4 W
Coll of William & Mary Lake Matoaka Amphitheater 89-E8; WE-A4 W
Coll of William & Mary Lake Matoaka Art Studio WE-A4 W
Coll of William & Mary Law Sch 89-H8 W
Coll of William & Mary Millington Hall WE-B4 W
Coll of William & Mary Monroe Hall WE-B3 W
Coll of William & Mary Morton Hall WE-A4 W
Coll of William & Mary Old Dominion Hall WE-B3 W
Coll of William & Mary Old Lodges WE-B3 W
Coll of William & Mary Phi Beta Kappa Mem Hall WE-B4 W
Coll of William & Mary Physical Lab WE-A4 W
Coll of William & Mary Physical Plant WE-C4 W
Coll of William & Mary Plumeri Baseball Pk 89-C4 J
Coll of William & Mary Presidents House WE-C3 W
Coll of William & Mary Randolph Complex WE-A3 W
Coll of William & Mary Residence Hall Complex WE-B4 W
Coll of William & Mary Rogers Hall WE-B4 W
Coll of William & Mary Student Hlth Ctr WE-A3 W
Coll of William & Mary Tercentenary Hall WE-B4 W
Coll of William & Mary Tucker Hall WE-B3 W
Coll of William & Mary Tyler Hall WE-B3 W
Coll of William & Mary Univ Ctr WE-B3 W
Coll of William & Mary Washington Hall WE-B3 W
Coll of William & Mary William & Mary Hall 89-E7 W
Coll of William & Mary Wren Bldg 89-G7; WE-C3 W
Coll of William & Mary Yates Hall WE-A3 W
Coll of William & Mary-Swern Library 89-F7 W
Commonwealth 158-H4 H
Hampton Univ 159-G7; HE-F6 H
Hampton Univ Admin Bldg HE-F7 H
Hampton Univ Alumni House HE-F6 H
Hampton Univ Armstrong Slater Trade Sch HE-F7 H
Hampton Univ B T Washington Mem HE-G6 H
Hampton Univ Bemis Lab HE-F8 H
Hampton Univ Buckman Hall HE-G6 H
Hampton Univ Clark HE-F7 H
Hampton Univ Comm HE-F6 H
Hampton Univ Convocation Ctr 159-G7; HE-H7 H
Hampton Univ Dubois HE-F8 H
Hampton Univ Early Childhood HE-G6 H
Hampton Univ Freeman Hall HE-G6 H
Hampton Univ Harkness Hall HE-F8 H
Hampton Univ Infirmary & Hu Bac HE-F6 H
Hampton Univ Katharine House HE-D7 H
Hampton Univ King Hall HE-G6 H
Hampton Univ Kittrell Hall HE-G6 H
Hampton Univ Marine Science HE-F8 H
Hampton Univ Math HE-F7 H
Hampton Univ Mus HE-E8 H
Hampton Univ Music HE-F7 H
Hampton Univ Natural Science HE-F7 H
Hampton Univ Olin Engineering HE-H6 H
Hampton Univ Phenix HE-F6 H
Hampton Univ Physical Education HE-G7 H
Hampton Univ Science echnology HE-F7 H
Hampton Univ Student Ctr HE-G7 H
Hampton Univ Whipple Barn HE-F6 H
Physical Education HE-G7 H
Riverside Hosp Sch of Nursing 141-C10 NN
Science Technology HE-F7 H
Thomas Nelson Comm Coll 142-C10; 158-D1 H
Virginia Inst of Marine Sci (Coll of Wllm & Mary) 92-J10; 108-J1; 109-A1 G
Virginia Inst of Marine Science Aquarium & Vis Ctr 109-A1 G
Virginia St Sch for the Blind 158-G10 H

COMMUNITY & RECREATION CENTERS

Hampton YMCA 159-C6 H
James City Williamsburg 89-C3 J
James River 107-B7 J
Kingsmill 106-E2 J
Langley AFB Youth Ctr 143-C9 L
Midtown 157-G4 NN
Newport News YMCA 157-H8 NN
North Phoebus 159-J6 H
Northampton 158-C4 H
Odd Road 126-K10 Pq
Old Hampton 159-D6; HE-B5 H
Ridley Circle 174-D4 NN
Salvation Army Corps 158-A2 H
Yorktown 109-B8 Y

FIRE COMPANIES

Abingdon Sta 3 & Hayes Res Sqd 92-J5 G
Abingdon Sta 5 92-G1 G

Charles City Co Dist 3 70-B8 CC
Fort Eustis Military Resv 123-H6 F
Hampton Police Academy & Fire Training Ctr 159-C4; HE-A1 H
Hampton Sta 1 159-D6; HE-B3 H
Hampton Sta 2 159-J8 H
Hampton Sta 3 174-J1 H
Hampton Sta 4 160-C4 H
Hampton Sta 5 144-C8 H
Hampton Sta 6 158-A4 H
Hampton Sta 7 159-H2 H
Hampton Sta 8 142-K6 H
Hampton Sta 9 158-E8 H
Hampton Sta 10 158-H2 H
Hampton Univ-Wig Wam 159-F7; HE-F7 H
James City Co 2 106-J3 J
James City Co 3 89-A10 J
James City Co 4 89-A1 J
James City-Bruton Co 1 56-A10 J
Langley AFB Co 1 143-E7 L
Langley AFB Co 3 143-C8 L
Newport News Sta 1 & Rescue 174-A3 NN
Newport News Sta 2 & Rescue 174-D3 NN
Newport News Sta 3 & Res Sqd 157-G5 NN
Newport News Sta 4 & Rescue 124-B2 NN
Newport News Sta 6 & Rescue 141-B4 NN
Newport News Sta 7 & Rescue 158-A9 NN
Newport News Sta 8 & Rescue 141-C9 NN
Newport News Sta 9 & Rescue 124-C10 NN
Newport News Sta 10 & Rescue 141-A9 NN
Newport News/Williamsburg Intl Airport 124-K9 NN
Poquoson 127-C10 Pq
US Naval Supply Ctr 91-A6 Y
Weir Creek Emergency Services 38-J1 NK
Williamsburg Hdq 89-G6; WE-C2 W
York Co 1 125-F5 Y
York Co 2 142-A1 Y
York Co 3 90-C8 Y
York Co 4 108-G4; YE-B5 Y
York Co 5 73-D4 Y
York Co 6 109-H8 Y
York Co Public Safety Bldg 125-E2 Y

GOLF COURSES

Deer Cove GC 91-F5 Y
Fords Colony GC 88-E2; 89-A3 J
Golden Horseshoe Gold Course 89-J9; WE-D4 W
Golden Horseshoe Gold Course Clubhouse WE-D4 W
Golden Horseshoe Green Course 89-K9; WE-F4 W
James River CC 140-J10; 156-K1; 157-A1 NN
Kiln Cr GC 141-C1 NN; 125-E10; 141-C1 Y
Kingsmill Bray Links GC 106-C5 J
Kingsmill Plantation GC 106-A4 J
Kingsmill River GC 106-C4 J
Kingsmill Woods GC 106-J4 J
Kiskiack GC 56-K8; 57-A8 J
Langley AFB Eaglewood GC 143-A6 L
Naval Base GC 175-K8 NBN
Newport News GC at Deer Run 124-F1 Y
Spotswood GC 89-J8 W
Stonehouse GC 39-K10; 55-J3; 56-A2 J
The Colonial GC 54-K8; 55-A8 J
The Hamptons GC 142-E10 H
The Pines GC 123-E8 F
The Woodlands GC 159-H6; HE-H5 H
Two Rivers CC 87-C10; 103-E1 J
US Naval Weapons Sta GC 92-B9 Y
Williamsburg CC 106-G2 Y
Williamsburg Natl GC 88-B8 J

HOSPITALS

Colonial 107-G8 NN
Eastern St 88-K4; 89-A3 J
Fort Monroe Health Clinic 160-A9 FM
Langley AFB 143-C9 L
Mary Immaculate 124-J7 NN
McDonald Army 123-J5 F
Newport News Gen 158-C10 NN
Newport News Hlth Ctr 174-D2 NN
Peninsula Behavioral Ctr 158-G2 H
Public Hlth Ctr 141-C10 NN
Radison Hotel HE-E5 H
Riverside 141-C10; 157-C1 NN
Sentara 159-F3 H
Sentara Careplex 158-H2 H
Sentara Hampton 159-A9 H
Sentara Hope Med Ctr 158-F2 H
Veterans Admin Med Ctr 159-G8; HE-H8 H
West Mercury Med Emergency 158-F4 H
Williamsburg Comm 89-E6; WE-A2 W

INFORMATION CENTERS

Colonial Williamsburg WE-C3 W
Colonial Williamsburg Vis Ctr 89-H6; WE-D1 W
Hampton Chamber of Commerce 142-G10 H
Hampton Vis Ctr 159-F7; HE-E5, E6 H
Hog Is WMA Vis Ctr 106-A10 SC
Jamestown Visitor Ctr 104-F7 J
Navy Base Tour 175-K10 NBN
Newport News Pk Campsite Information 108-A10 NN
Newport News Vis Ctr & Pk Hdq 124-B2 NN
Riverview Farm Pk Vis Ctr 140-E7 NN
Virginia Inst of Marine Science Aquarium & Vis Ctr 109-A1 G
Williamsburg Area Convention & Vis Bureau 90-A7; WE-F3 W
York River St Pk Taskinas Pt Vis Ctr 57-F4 J
Yorktown Vis Ctr & Pk Hdq 108-K4; YE-E3 Y

ISLANDS & POINTS

Barren Pt 58-E6 G
Barrets Pt 103-A2 J
Bay Pt 126-K6 Pq
Bay Tree Pt 110-H10 Y
Big Marsh Pt 70-E4 J
Black Pt 105-C8 J
Black Snake Is 127-C5 Pq
Blunt Pt 156-G2 NN
Candy Is 173-B10 I
Carmines Islands 92-E5 G
Catlett Islands 92-B3 G
Cedar Pt 143-A2 Pq
Church Pt 104-E7 J
Cow Is 127-A5 Pq
Cowpen Neck 58-J8 G
Crispy Pt 139-J7 F
Curtis Pt 140-B8 F
Dandy Pt 144-C5 H
Days Pt 155-H7 I
Drum Pt 127-G6 Pq
East Is 173-A7 I
Ferry Pt 86-K9 CC; 74-E2 Y
Fishing Pt 173-A6 I
Gaines Pt 109-G1 G
Gas Dock Pt 54-F10 J
Glass House Pt 104-D6 J
Gloucester Pt 108-K1 G
Goodwin Islands 110-E6 Y
Goodwin Pt 155-J10 I
Goose Is 123-B5 F
Gordon Is 87-B7 J
Green Pt 92-A3 G
Gross Pt 173-B7 I
Grunland Pt 144-D4 H
Hammock Pt 128-A8 Pq
Hicks Is 54-H7 J
Hog Is 105-K9; 122-A1 SC
Hog Pt 106-A9 SC
Hunts Pt 126-F5 Pq
Jail Pt 140-B8 F
Jamestown Is 104-H8; 105-A9 J
Lower Pt 104-K10 J
Marsh Is 127-A6 Pq
Marsh Pt 143-B3 L; 127-G4 Pq
Marshy Pt 139-C2 F
Messick Pt 143-K3 Pq
Mulberry Is 123-B10; 139-H3 F; 122-K10 NN
Mulberry Pt 122-J10 NN
Mumfort Islands 92-H7 G
Newport News Pt 174-E7 NN
Northend Pt 144-D3 H
Oak Is 143-C2 Pq
Old Neck 70-J2 CC
Oyster Is 127-G7 Pq
Parsons Is 70-J6 CC
Penniman Spit 91-H6 Y
Plumtree Is 128-C10; 144-D1 Pq
Plumtree Pt 128-D10 Pq
Point of Rocks 109-A4; YE-F4 Y
Poley Pt 91-G7 Y
Purtan Is 58-D3 G
Pyping Pt 104-H6 J
Quarter Pt 109-D1 G
Ragged Is 173-A8 I
Sachems Head 74-C1 J
Sandy Pt 92-B8 Y
Sewells Pt 175-J6 NBN
Sheep Is 127-C6 Pq
Shields Pt 87-A2 J
Ship Pt 126-E3 Y
Simpson Is 71-B10 J
Stony Pt 143-J5 H; 92-E10 Y
Swanns Pt 104-B8 SC
Swash Hole Is 139-C3 F
Tabbs Pt 143-D4 L
Terrapin Pt 40-J3 NK
Thorofare Is 139-K3 F
Thorofare Pt 139-K3 NN
Tin Shell Pt 143-F4 Pq
Tue Pt 110-J4 Y
Walnut Pt 122-D3 SC
Ware Stick Pt 127-J8 Pq
Watts Pt 70-E1 CC
Whalebone Is 127-H7 Pq
Wilcox Neck 54-D8 CC
Willoughby Pt 143-G6 L
Windmill Pt 144-A5 H
Wright Is 71-A7 J
Yarmouth Is 71-B9 J
York Pt 126-G2 Y
York River Cliffs 108-G2; YE-A1 Y

LAKES & STREAMS

Adams Cr 58-B1 G
Ajacan Lake 105-E4 J
Back Cove 143-J3 Pq
Back Cr 54-H7 J; 109-J7; 110-C8 Y
Back River 104-F6 J
Back River Marsh 104-H6 J
Back River Northwest Br 142-K3; 143-C3 L; 142-K3; 143-C3 Pq
Back River Southwest Br 143-C10; 159-C2 H; 143-C10 L
Bailey Cr 123-E3 F
Bakers Cr 40-K1; 41-A1 KQ
Ballard Cr 108-E2 Y
Balls Pond 38-J3 NK
Baptist Run 108-D6 Y
Barlows Pond 73-D1 J; 73-D1 Y
Barnes Swamp 39-B9 J; 39-B9 NK
Barrows Cr 54-B9; 70-C2 CC
Bay Tree Cr 110-G10 Y
Beaver Pond 104-D1 J
Beaverdam Cr 38-E2 NK; 108-G9 Y
Beaverdam Pond 74-H9 Y
Beaverdam Swamp 74-F10 Y
Bells Oyster Gut 144-A2 Pq
Bennett Cr 127-A7 Pq
Big Bethel Res 141-K5 H; 141-K5 NN; 141-K5 Y
Big Salt Marsh 127-F7; 143-H1 Pq
Bigler Mill Pond 74-G7 Y
Billy Wood Canal 158-E1 H
Bird Swamp 56-B2 J
Black Duck Gut 104-A8 SC
Black Swamp 91-G10; 107-D1 Y
Blackstump Cr 71-C10; 87-B1 J
Bland Cr 58-G5 G
Blows Cr 123-B8 F
Blows Mill Run 107-E3 J; 107-E3 Y
Boathouse Cr 126-D2 Y
Brick Kiln Cr 142-F5 H; 141-E3 NN; 142-F5 Pq; 142-F5 Y
Brights Cr 159-E5; HE-B3, D1 H
Broad Swamp 103-A8 SC
Browns Lake 123-F7 F
Bushneck Cr 87-D5 J
Butlers Gut 123-F8 F
Buzzard Bay 87-C4 J
Cabin Cr 126-F1 Y
Carter Cr 73-G5; 74-C6 Y
Cedar Cr 142-K1; 143-A2 Pq
Cedarbush Cr 92-A1 G
Cheatham Pond 90-K4; 91-A3 Y
Chesapeake Bay 160-F8 FM; 144-F3; 160-F8 H; 126-J1; 128-B5 Pq; 110-H6; 126-J1 Y
Chickahominy River 54-B7; 70-F4; 86-K2; 87-A2, A9 CC; 54-B7; 70-F4; 71-A5; 86-K2; 87-A2, A9; 103-A1 J; 54-B7 NK
Chisel Run 88-J4; 89-A2 J
Chisman Cr 126-A1 Y
Church Cr 159-C10 H
Claxton Cr 110-E8 Y
Colby Swamp 71-K10; 87-K1 J
Coleman Swamp 92-K4 G
College Cr 89-D5; 105-G2 J; 89-D5, F9; 105-G2 W
Cooper Cr 173-A7 I
Cow Swamp 56-G3 J
Cranstons Pond 72-A5 J
Cross Cr 103-E10 SC
Croudes Swamp 88-D4 J
Dardanelles Pond 74-E6 Y
Davis Pond 39-J6 NK
Deep Cr 140-G7 NN
Deer Lake 72-D6 J
Diascund Cr 54-G6 J; 54-G6 NK
Diascund Cr Res 54-B1 J; 38-B9, G9; 39-A8, A10; 54-B1 NK
Drum Island Flats 127-H5 Pq
Easton Cove 127-B8 Pq
Edwards Swamp 55-D7 J
Elizabeth River 175-D7 N; 175-D7 NBN; 175-D7 P
Eustis Lake 123-D5 F
Felgates Cr 91-G7; 108-A1 Y
Fire Pine Cr 127-J10 Pq
Fishers Cr 140-F10 NN
Flat Gut 144-B2 Pq
Floods Hole 144-F5 H
Floyds Bay 127-A8 Pq
Fore Landing Cr 143-F1 Pq
Fort Cr 123-A9 F
Fox Cr 58-H8 G
France Swamp 56-D8, E3 J
Front Cove 143-J2 Pq
Gable Br 58-B2 G
Glebe Gut 105-C6 J
Goddins Pond 40-B4 NK
Goose Cr 110-B10; 126-B1 Y
Gordon Cr 87-B8, F5 J
Grays Cr 104-B10 SC
Great Run 108-F4; YE-A5 Y
Gressitt Pond 41-E1 KQ
Grices Run 107-A8 J
Grove Cr 106-H4 J
Grunland Cr 144-E7 H
Gum Hammock Cr 128-B10 Pq
Guthrie Cr 41-H6 KQ
Halfway Cr 105-H3 J
Hampton River 159-F5; HE-D6, E7, F5 H
Hampton Roads 159-F10; 175-E5 H; 173-F8 I; 175-E5 N; 175-E5 NBN; 173-F8; 174-C8 NN
Haring Swamp 74-D9; 90-D1 Y
Harris River 144-A6 H
Harwoods Mill Res 125-D6 Y
Hawkins Pond 144-G7 H
Haystack Gut 104-A9 SC
Herberts Cr 159-F3 H
High Cedar Cr 143-J1 Pq
Hipps Pond 91-B7 Y
Hockley Cr 41-C1 KQ
Hodges Cove 126-D3 Y
Hog Island Cr 122-B3 SC
Hog Neck Cr 71-A2 J
Homewood Cr 122-B1 SC
Hunnicut Cr 122-A7 SC
Indian Cr 57-K3 G
Indian Field Cr 92-A9 Y
Indian River 159-A10 H
Indigo Lake 156-H1 NN
Island Cr 123-E6 F
Jail Cr 140-A7 F
James River 139-B4 F; 159-F10 H; 122-E5; 155-D2; 173-E3 I; 103-B4; 104-D8; 105-J6; 107-A9; 122-E5; 123-A6 J; 122-E5; 123-A6; 156-C4; 157-B7; 173-E3 NN; 103-B4; 104-D8; 105-J6; 122-E5 SC
Joachim Lake 105-E3 J
Johns Cr 159-H8 H
Jolly Pond 87-J2 J
Jones Cr 159-F8; HE-E8 H
Jones Pond 90-H6 Y
Jones Run 124-C3 NN
Kettle Pond 157-B3 NN
King Cr 90-J9; 91-A8, E6 Y
Kingsmill Cr 104-J7 J
Kingsmill Pond 106-D3 J
Kitchums Pond 103-J1 J
Lake Corbin 140-K9 NN
Lake Hampton 159-A5 H
Lake Loring 105-B1 J
Lake Matoaka 89-D8; WE-A4 W
Lake Maury 141-D10; 157-C4 NN
Lake Nice 56-E8 J
Lake Norvell 58-A9 J
Lake Pasbehegh 104-A2 J
Lake Powell 104-K2; 105-A2 J
Lake Queen Anne 156-G1 NN
Lake Tormentor 155-B6 I
Lambs Cr 126-D8 Pq; 126-D8 Y
Lawnes Cr 122-B10 I; 122-B10 SC
Lee Hall Res 107-K10; 123-K2; 124-B1 NN; 108-B10, E10 Y
Lee Pond 108-A1 Y
Leigh Cr 58-E4 G
Little Cr 71-E6 J
Little Cr Res 55-J10; 71-F4 J
Lloyd Bay 127-C5 Pq
Long Cr 144-E9 H; 143-G3 Pq
Long Hill Swamp 72-J10; 73-A9; 88-H2 J
Lucas Cr 140-B2 NN
Lynnhaven Lake 158-E3 H
Lyons Cr 126-J7 Pq
Meadow Swamp 41-K2 KQ
Mill Cr 160-A8 FM; 160-A8 H; 54-H8; 55-A8, G8; 88-K9; 104-K1; 105-B6 J
Mill Cr Pond 89-A8 J
Milstead Cr 123-D5 F
Moores Cr 126-B9 Y
Morleys Gut 123-A9 F
Morris Bay 41-J9 G
Morris Cr 70-A10; 86-G3 CC
Morrisons Cr 123-D10; 139-D2 F
Nayses Bay 87-E6 J
Nettles Cr 86-K5; 87-A5 J
Newmarket Cr 141-J10; 157-H3; 158-E7; 159-A5 H; 141-J10; 157-H3 NN
Newport News Cr 174-D5 NN
Northwest Br 103-K1 J
Northwest Br Back River 142-K3; 143-C3 L; 142-K3; 143-C3 Pq
Northwest Br Shellbank Cr 87-J10 J
Old Mill Pond 73-A1 J; 73-A1 Y
Old Neck Cr 70-H4, H5 CC
Pagan River 155-C10, H10 I
Paper Mill Cr 89-G10 W
Parsons Cr 70-H5 CC
Passmore Cr 104-J8; 105-A9 J
Pates Cr 105-E5 J
Patricks Cr 126-A5 Y
Pauly Run 139-G3 F
Penniman Lake 91-D6 Y
Philbates Cr 40-D3 NK
Phillips Lake 159-H2 H
Pine Woods Pond 87-H7 J
Pitch and Tar Swamp 104-H8 J
Point Swamp 87-D1 J
Pond No 10 91-D8 Y
Pond No 11 91-C8 Y
Pond No 11A 91-B8 Y
Pond No 12 91-B8 Y
Poplar Cr 92-B2 G
Poquoson Flats 127-G1 Pq
Poquoson River 126-E5 Pq; 125-C3, J9; 126-E5 Y
Poropotank Bay 41-H9 G
Powell Lake 74-D3 Y
Powhatan Cr 88-E4, E8; 104-H3 J
Purtan Bay 58-D5 G
Purtan Cr 58-D3 G
Quarter March Cr 126-A7 Y
Queen Cr 89-K3 W; 73-C6; 89-K3; 90-E2; 91-A2 Y
Queens Lake 90-D4 Y
Ragged Island Cr 173-A9 I
Rhine River 106-H4 J
Richardson Millpond 40-C10; 56-C1 J; 40-C10; 56-C1 NK
Richardson Swamp 39-H7 NK
Roberts Cr 126-G6 Pq
Robinson Cr 174-J1 H
Rock Cr 127-D4 Pq
Roosevelt Pond 92-C10; 108-C1 Y
Salt Pond 160-D2 H
Salters Cr HE-C6, D6 H; 174-G2 NN
Sandy Bay 104-E6 J; 127-B4 Pq
Sandy Cr 58-K10; 74-K1 G
Scotts Pond 72-K9; 73-A9 J
Shellbank Cr 103-J1 J
Shellbank Cr NW Br 87-J10 J
Shipyard Cr 71-A7 J
Skiffes Cr 123-E2 F; 107-C6; 123-E2 J; 123-E2 NN
Skiffes Cr Res 107-F8 J; 107-F8 NN
Skimino Cr 72-K4; 73-A3, F2; 74-A1 J; 72-K4; 73-A3, F2; 74-A1 Y
Skimino Pond 74-C3 Y
Sluice Millpond 140-H3 NN
Southwest Br Back River 143-C10; 159-C2 H; 143-C10 L
Stoney Run 124-A8, F7 NN
Sunken Marsh 70-H4 CC
Sunset Cr 159-D8; HE-B8, C8 H
Swash Hole 139-D3 F
Tabbs Cr 142-K5; 143-B4 L
Taskinas Cr 57-B5 J
Taylor Pond 39-J1 NK
The Thorofare 104-J6; 105-A7 J
Thorofare Cr 139-K3 F; 139-K3 NN; 110-D6 Y
Thorofare Cr 128-C10 Pq
Tide Mill Cr 143-A10; 158-J2; 159-A1 H
Tide Mill Pond 158-K1; 159-A1 H
Timber Swamp 38-C7 NK
Timberneck Cr 92-E3 G
Tomahund Cr 86-A5, D8 CC
Topping Cr 127-B10; 143-B1 Pq
Tormentor Cr 155-C9 I
Tutters Neck Pond 90-A10 W
Uncles Neck Cr 71-A3 J
Upper Back Cr 54-H6 J
Wahrani Swamp 38-J10; 39-A8 NK
Wallace Cr 144-D5 H
Waller Mill Res 73-F8; 89-F2 Y
Warburton Pond 87-J5 J
Ware Cr 39-J8; 40-H10; 56-F1 J; 39-J8; 40-H10; 56-F1 NK
Warehams Pond 106-E5 J
Warwick River 123-J8; 139-J1 F; 123-J8; 139-J1; 140-A4 NN
Watts Cr 143-D2 Pq
West Br Wormley Cr 109-B6 Y
White House Cove 126-J8 Pq
White Pond Swamp 144-E10 H
Whiteman Swamp 90-G10 Y
Whittaker Lake 103-G1 J
Willett Cove 126-E2 Y
Williams Cr 155-E8 I
Winders Pond 125-H5 Y
Wingfield Lake 87-J10 J
Wood Cr 107-C10; 123-C1 J
Wormley Cr 109-E7 Y
Wormley Cr W Br 109-B6 Y
Wormley Pond 109-A6 Y
Yarmouth Cr 71-D8; 72-D5 J
Yoder Pond 140-G5 NN
York River 57-E2; 58-C6; 74-F3; 91-E2; 92-D6; 108-H1; 109-D2; 110-E3 G; 41-B5; 57-E2; 58-C6 J; 40-G2; 41-B5; 57-E2 KQ; 40-G2; 41-B5 NK; 58-C6; 74-F3; 91-E2; 92-D6; 108-H1; 109-D2; 110-E3; YE-E1 Y
Yorktown Cr 108-H2; YE-A1 Y

LIBRARIES

Coll of William & Mary-Swern 89-F7; WE-A4 W
Fort Eustis Military Resv 123-H5 F
Hampton 159-D7; HE-B7, E7 H
Hampton Univ 159-F7; HE-E7 H
James City Co 56-H10 J
Lee Hall 107-J9 NN
Main Street 157-G5 NN
Northampton 158-B3 H
Outreach & Ext Services 157-G5 NN
Pearl Bailey 174-E3 NN
Phoebus 159-J8 H
Poquoson 142-G1 Pq
Rockefeller WE-E2 W
Tabb 141-J2 Y
Virgil Grissom 124-C10 NN
West Avenue 174-A4 NN
Williamsburg Reg 89-G7; WE-C3 W
Willow Oaks 159-H1 H
York Co 109-B9 Y

MARINAS & RAMPS

Amorys Wharf Ramp 143-G4 Pq
Back Cr Pk Ramp 109-K7 Y
Back River 144-C5 H
Back River Ramp 143-J3 Pq
Bennetts Cr #2 Ramp 127-B8 Pq
Bluewater Yacht Yard Ramp 144-D6 H
Buckroe Beach Fishing Pier 160-D5 H
Busch 106-D5 J
Capahosic Ramp 58-J10 G
Chickahominy 70-J2 J
Chickahominy River Ramp 70-K1 J
Chisman Cr 126-C1 Y
Colonial Harbor 54-D6 NK
Commonwealth of Virginia Fuel Pier 91-G5 Y
Croaker Ramp 57-D2 J
Dandy Haven 144-B5 H
Dandy Pt Ramp 144-D5 H
Denbigh Landing Ramp 139-K1 NN
Diascund Cr Res Ramp 54-F1 NK
Field Landing Ramp 92-F2 G
Gloucester Pt Pk Fishing Pier 108-K1 G
Gloucester Pt Ramp 108-J1 G
Gosnolds Hope Pk Ramp 143-H8 H
Grandview Fishing Pier 144-F9 H
Hampton 159-E7; HE-D7 H
Hampton Boat Tours 159-F7; HE-E6 H
Hampton Public Boat Dock HE-E6 H
Hampton Roads Marine Corp 159-E8 H
Hampton Yacht Club 159-E7; HE-D7 H
Harbour Ctr Bldg HE-E5 H
Harwoods Mill Fishing Concession Ramp 125-D7 Y
Hideaway 70-D2 CC
Hunts Ramp 126-F6 Pq
Islander 126-K8 Pq
James River 140-F8 NN
James River Fishing Pier 157-G8 NN
Jamestown 104-J4 J
Jamestown Yacht Basin 104-F4 J
Jones 159-F7; HE-D7 H
Joys 159-G7; HE-E5 H
Langley 143-E6 L
Langley AFB Yacht Club 143-F8 L
Lawson & Son 159-F8 H
Leeward Mun 157-G9 NN
Marina Cove Boat Basin 143-K6 H
Marina Pooles Grant 159-F6; HE-F3 H
McGurl Yacht Ramp 159-E8 H
Menchville 140-G7 NN
Morris Cr Ramp 86-G2 CC
Naval Base Norfolk Pier 2 175-G9 NBN
Naval Base Norfolk Pier 3 175-G9 NBN
Naval Base Norfolk Pier 5 175-G9 NBN
Naval Base Norfolk Pier 6 175-G8 NBN
Naval Base Norfolk Pier 7 175-G8 NBN
Naval Base Norfolk Pier 10 175-H8 NBN
Naval Base Norfolk Pier 11 175-H7 NBN
Naval Base Norfolk Pier 12 175-H7 NBN
Naval Base Norfolk Pier 23 175-G10 NBN
Naval Base Norfolk Pier 24 175-G10 NBN
Naval Base Norfolk Pier 25 175-G10 NBN
Naval Base Norfolk Pier C 175-H8 NBN
Naval Base Norfolk Pier D 175-G8 NBN
Naval Base Norfolk Pier E 175-G8 NBN
Naval Base Norfolk Pier F 175-G8 NBN
Naval Base Norfolk Pier G 175-G8 NBN
Naval Base Norfolk Pier H 175-H8 NBN
New Old Pt Comfort 159-K10 FM
Newport News Cruise Terminal 174-A5 NN
Newport News Pk Boat Rental 124-C1 NN
Newport News Pk Ramp 108-A10 NN
Newport News Tour Boat 174-D5 NN
Peterson Yacht Basin Ramp 174-F3 NN
Pier 2 173-K4 NN
Pier 4 173-K3 NN
Pier 5 173-K3 NN
Pier 6 173-K3 NN
Pier 9 173-J2 NN
Pier 442 123-C4 F
Poquoson 126-K8 Pq
Powhatan Cr Canoe Access 104-H2 J
Public Boat Dock HE-E6 H
Public Landing 143-B2 Pq
Public Landing Ramp 157-G8 NN
Queens Lake Marina Corp 90-F3 Y
R M Mills 110-B8 Y
Raven 159-E8 H
Rodgers A Smith Landing Ramp 126-A9 Y

Salt Pond 160-D2 H
Sims 70-J2 J
Smiths Marine Railway 126-B1 Y
Southall 160-E1 H
Sunset 159-E8 H
Sunset Ramp 159-D8 H
The Wharf 108-J2; YE-D1 Y
Thomas 126-E1 Y
US Naval Supply Ctr Supply Pier 91-G4 Y
Wallaces 144-C5 H
Warwick Yacht Club 140-G7 NN
Wildey Marine 126-E2 Y
Wormley Cr 109-E6 Y
Wormley Cr Ramp 109-E8 Y
Yorktown Town Pier 108-K3 Y

MILITARY & FEDERAL FEATURES

Camp Peary Landing Strip 74-J10 Y
CEBAF Ctr 141-C5 NN
Department of Def Armed Forces Exp Tng-Camp Peary 58-A10 J; 74-B4; 90-D1; 91-A2 Y
Fort Eustis Military Resv 123-C8; 139-E2 F; 123-C8 J; 122-K10 NN
Fort Monroe 160-C8 FM
Fort Monroe Chapel of the Centurian 160-A10 FM
Fort Monroe Commissary 159-K10 FM
Fort Monroe East Gate 160-B10 FM
Fort Monroe Engineer wharf 160-A10 FM
Fort Monroe Main Casemate Mus 160-A10 FM
Fort Monroe Main Gate 160-A10 FM
Fort Monroe North Gate 160-A9 FM
Fort Monroe Officers Club 160-C8 FM
Fort Monroe Old Pt Lighthouse 160-A10 FM
Fort Monroe Post Exchange 160-B10 FM
General Services Admin Bldg 125-F1 Y
Hampton Armory 158-H2 H
Langley Air Force Base 143-B6 H; 142-K5, K8 L
Langley Air Force Base Commissary 143-B9 L
Langley Air Force Base Enlisted Club 143-D5 L
Langley Air Force Base King Street Gate 143-E10 H
Langley Air Force Base LaSalle Gate (Main Gate) 143-B10 H
Langley Air Force Base Officers Club 143-E9 L
Langley Air Force Base Post Exchange 143-C9 L
Langley Air Force Base West Gate 142-K8 H
Marine Corps Reserve Training Ctr 157-H9 NN
NASA Hdq 142-J6 H
NASA Research Ctr 142-H5 H; 143-A6 L
NASA Research Ctr Durand Gate 142-K6 H
NASA Research Ctr Main Gate 4 142-K7 H
NASA Research Ctr Wythe Cr Gate 142-H5 H
National Pk Service Maintenance 108-G5 Y
Naval Base Norfolk 175-J8 NBN
Naval Base Norfolk Commissary 175-K9 NBN
Naval Base Norfolk Dental Clinic 175-K10 NBN
Naval Base Norfolk Exchange 175-K10 NBN
Naval Base Norfolk Gate 1 175-J10 NBN
Naval Base Norfolk Gate 2 175-K9 NBN
Naval Base Norfolk Medical Clinic 175-K10 NBN
Thomas Jefferson Natl Accelerator Facility 141-C4 NN
US Coast Guard Reserve Training Ctr 109-D6 Y
US Military Resv 158-H2 H
US Naval Supply Ctr 91-A5; 109-B6 Y
US Naval Supply Ctr Hdq 91-D5 Y
US Naval Weapons Sta 107-D3 J; 107-D3 NN; 90-G7, K9; 91-D9; 106-K1; 107-D3; 108-A2 Y
US Naval Weapons Sta Main Gate 108-C3 Y

PARK & RIDE

Hayes 92-J6 G
Lightfoot Road 73-C5 Y
Old Courthouse Road 124-C9 NN
Route 30 56-H9 J
Yorktown Road 107-H7 NN

PARKS & RECREATION

Aberdeen Bog Park 158-E6 H
Abingdon Park 92-H4 G
Air Power Park & Museum 159-A4 H
Anderson Park 174-G4 NN
Armstrong Athletic Field HE-G7 H
Back Cr Pk 109-K8 Y
Beechlake Pk 141-J7 NN
Bicentennial Pk 89-H8; WE-C4 W
Big Bethel Rec Area 142-B6 H
Bluebird Gap Farm 158-H5 H
Briarfield Pk 158-E8 H
Buckroe Beach 160-D4 H
Buckroe Pk 160-D4 H
Busch Gdns 106-G3 J
Carters Grove 106-K7; 107-A7 J
Carters Grove Reception Ctr 106-K7 J
Central Pk 142-G10 H
Charles E Brown Pk 108-B4 Y
Chickahominy WMA 70-D8; 86-B3, E1 CC
Chisman Cr Pk 125-G2 Y
Christopher Newport Pk 174-A4 NN
Coll of William & Mary Adair Gymnasium WE-A4 W
Coll of William & Mary Barksdale Playing Field WE-B4 W
Coll of William & Mary Baseball Field WE-B3 W
Coll of William & Mary Busch Soccer Field WE-A3 W
Coll of William & Mary Busch Tennis Court WE-A3 W
Coll of William & Mary Cary Field at Zable Stadium WE-B3 W
Coll of William & Mary Intramural Field WE-A2 W
Coll of William & Mary Natl Wildflower Refuge WE-B3 W
Coll of William & Mary Recreational Sport Ctr WE-A3 W
Coll of William & Mary Sunken Garden WE-B3 W
College Landing Pk 89-F10 W
Colonial Natl Hist Pk 104-H7; 105-A8 J; 103-H8 SC; 91-E7; 108-E6; 109-A5; YE-A2 Y
Colonial Natl Hist Pk (Greensprings Plantation) 88-A9 J
Colonial Natl Hist Pk Hdq & Yorktown Ctr 108-K4 Y
Colonial Natl Hist Pk Hdq & Yorktown Vis Ctr YE-E3 Y
Colonial Natl Hist Pk Service Maintenance 108-G5 Y
Colonial Williamsburg WE-D3 W
Darling Stadium 159-D7; HE-A7 H
Deer Pk 141-D10 NN
Denbigh Pk 139-K1 NN
Eason Pk 159-C7; HE-A7 H
Fantasy Farm 140-F6 NN
Fort Boykin Hist Pk 155-A5 I
General Matthew Ridway Pk 159-F3 H
Gloucester Pt Pk 108-K1 G
Gloucester Pt Pk Fishing Pier 108-K1 G
Gosnolds Hope Pk 143-G9 H
Grandview Nature Preserve 144-E5 H
Greensprings Plantation Natl Hist Site 88-A9 J
Grunland Pk 144-D5 H
Hampton Carousel Pk Plaza HE-D5 H
Hampton Coliseum 158-J5 H
Hampton Univ Armstrong Athletic Field HE-G7 H
Harwoods Mill Mountain Bike Tr 125-C6 Y
Hog Is Game Refuge 122-A2, A7 SC
Hog Is WMA 106-A9 SC
Hog Is WMA Viewing Tower 106-A10 SC
Huntington Pk 157-G8 NN
Huntington Pk Beach 157-G8 NN
Ivy Farms Pk 157-H2 NN
James City Co Dist Pk 72-B9; 88-B1 J
James City Co Dist Pk Sports Complex 72-G8 J
Joynes Terr Pk 158-D7 H
Kiln Cr Pk 141-F1 Y
King Lincoln Pk 174-E6 NN
King Lincoln Pk Interpretive Ctr 174-D5 NN
Kiwanis Pk 89-C4 W
Langley Speedway 142-H7 H
Lees Mill Pk 123-K4 NN
Little Cr Res Rec Area 71-E2 J
Mariners Museum Pk 141-C10; 157-B2 NN
Mid Co Pk 88-J8 J
Mill Pt Pk 159-F6; HE-E5 H
Mitchell Pk HE-C5 H
Monitor-Merrimac Overlook Pk 174-G3 NN
Municipal Pk 127-A10 Pq
Murphy Field 123-H5 F
National Pk Service Cheatham Pond Area 90-H4; 91-A3 Y
Newport News Pk 108-A9; 124-D2, F3 NN; 108-A9; 109-A10; 124-D2; 125-C6 Y
Newport News Pk Archery Range 108-A10 NN
Newport News Pk Boat Rentals 108-A10 NN
Nicewood Pk 140-E3 NN
Oak Avenue Minipark 174-E2 NN
Peninsula Boy Scout Resv 72-A5 J
Phillips Pk 126-K10 Pq
Pipsico Resv (Boy Scouts of America) 103-A9 SC
Plum Tree Is Natl Wildlife Refuge 126-K5; 127-A5, F8; 144-A1 Pq
Poquoson Mun Pk 126-K10 Pq
Poquoson Parks & Rec Hdq & Mun Comm Ctr 126-K9 Pq
Potters Field Pk 124-C9 NN
Quarterpath Pk 90-A8 W
Ragged Island WMA 173-A8 I
River Drive Beach 157-H9 NN
Riverview Farm Pk 140-E6, F7 NN
Robinson Pk 174-K1 H
Sam Houston Pk 143-J10; 159-J1 H
Sandy Bottom Nature Pk 141-K9; 158-A1 H
South Lawson Pk 127-D10; 143-C1 Pq
Teardrop Pk 157-E1 NN
Tindalls Pt Pk 92-J10; 108-J1 G
Todd Stadium 141-A9 NN
Upper Co Pk 55-G4 J
Waller Mill Pk 73-F9; 89-F2 Y
War Mem Stadium 158-J9 H
Water Country USA 90-H9 Y
Wolf Trap Pk 125-F3 Y
Y H Thomas Pk 159-C4; HE-A1 H
York Co New Quarter Pk 90-G3 Y
York River St Pk 57-D4 J
Yorktown (Little League Pk) 109-B8 Y
Yorktown Beach 108-H2, J2; YE-C1, D1 Y

PLACES OF WORSHIP

Adath Jeshurun 141-A7 NN
Aldersgate United Meth 158-K10 H
Asbury United Meth 124-B5 NN
Ascension of Our Lord Byzantine 89-F5 Y
Beacon Bapt 158-H1 H
Bethel AME HE-C4 H
Bethel Bapt 125-K10 Y
Bethel Temple Assembly of God 158-E4 H
Bible Bapt 127-B10 Pq
Bnai Israel 159-A9 H
Bruton Parish 89-H7 W
Carver Mem Presb 174-D3 NN
Carys 142-D1 Y
Chapel No 4 142-C5 Y
Chickahominy Bapt 71-J4 J
Christian Life Ctr 72-H10 J; 125-K10 Y
Christian Science 89-E8; WE-B4 W
Church of God 141-K3 Y
Church of the Lord Jesus Christ 174-F2 NN
Church of the Nazarene 90-D8 Y
Colonial Chapel 89-C3 W
Colonial Pl Ch of Christ 159-A9 H
Colossian Bapt 124-F2 NN
Dandy Bapt 110-A6 Y
David Adams Mem Chapel 175-K9 NBN
Denbigh Presb 140-C1 NN
Ebenezer 54-C3 NK
Emmaus Bapt 126-G8 Pq
Faith House Worship Ctr 86-J9 CC
First Bapt 140-K6 NN; WE-B3 W
First Bapt East End 174-C3 NN
First Bapt of Denbigh 140-G1 NN
First Calvary Bapt 160-B1 H
First Ch of Newport News Bapt 174-E3 NN
First Chr 158-C4 H
First Friends 158-B2 H
First Meth HE-E5 H
First Presb 159-E7; HE-B7 H
Foxhill Meth 160-C1 H
Full Gospel Ch of Deliverance 174-A3 NN
Gethsemane Bapt 174-D2 NN
Gloucester Ch of Christ 92-K6 G
Good Hope 38-A4 NK
Grace Bapt 90-D8 J
Grace Epis 108-J3; YE-D2 Y
Grafton Bapt 125-F5 Y
Grafton Chr 125-D3 Y
Greensprings 88-H8 J
Hampton Bapt 159-E6; HE-D4 H
Harbor Bapt 142-B6 H
Hickory Neck Epis 56-B8 J
Hidenwood Presb 140-J10 NN
Holy Tabernacle Ch of Deliverance 124-D9 NN
Hornsbyville Meth 109-E9 Y
Immaculate Conception 158-H3 H
Ivy Bapt 174-H2 NN
Ivy Mem Bapt 158-H2 H
James River Bapt 88-B2 J
Jamestown Presb 88-G10 J
King of Glory Luth 88-K2 J
King Street Bapt HE-D4 H
Kingdom 142-B6 H
Kingdom Hall 158-K9; 159-F1 H; 73-B9 J; 124-C10; 174-B2 NN
Kingdom Hall of Jehovahs Witnesses 125-J4 Y
Kirkwood Presb 141-J2 Y
Lakeside Ch of God 141-H6 NN
Langley AFB Main Base Chapel 143-E8 L
LDS 124-J6 NN
Lebanon 107-H6 NN
Lee Hall Bapt 107-H8 NN
Liberty Bapt 158-A3 H; 54-G2 NK
Lincoln Pk Bapt 159-B7 H
Little England Chapel 159-D8 H
Little Zion Bapt 158-H5 H; 106-K5 J
Macedonia Bapt 174-G2 NN
Magruder 74-B9 Y
Maranatha Bapt 141-H1 Y
Marquand Mem HE-E7 H
Morning Star 142-H10 H
Morning Star Bapt 107-D7 J
Mount Ararat Bapt 89-J7; WE-E3 W
Mount Gilead Bapt 107-A5 J
Mount Nebo 39-E3 NK
Mount Olive 40-D4 NK
Mount Pilgrims 73-G5 Y
Mount Pleasant Bapt 70-A2 CC
Mount Zion Bapt 70-A8 CC
New Bethel 142-A2 Y
New Life Ch of God 92-J2 G
New Mount Olive Bapt 158-D7 H
New Zion Bapt 72-F10 J
North Riverside Bapt 157-A1 NN
North Side Ch of Christ 141-F5 NN
Oak Grove 89-H3 Y
Olive Branch 56-D10 J
Olivet Chr 124-C9 NN
Our Lady of Victory 175-K9 NBN
Our Redeemer Luth 142-B1 Y
Our Savior 72-F1 J
Poquoson Assembly of God 126-H10 Pq
Poquoson Bapt 142-G2 Pq
Providence 125-K4 Y
Queen Street Bapt 159-E6; HE-C5 H
Resurrection Luth 141-F7 NN
Rising Sun 108-C3 Y
Rodef Shalom 158-A6 H
Saint Augustine Cath 174-D3 NN
Saint Bede Cath 89-F6; WE-B2 W
Saint Jeromes Cath 140-A1 NN
Saint Joan of Arc Cath 109-A9 Y
Saint Johns HE-F6 H
Saint Johns Bapt 90-F9 Y
Saint Johns Epis HE-C5 H
Saint Lukes Meth 125-H8 Y
Saint Mark Luth 125-D2 Y
Saint Martins Epis 89-B10 J
Saint Marys 159-K10 FM
Saint Mathews Anglican 157-G5 NN
Saint Matthew Reform AME 174-E3 NN
Saint Paul AME 174-F3 NN
Saint Roses Cath 158-J10 H
Saint Steven Luth 89-E8; WE-B4 W
Saint Vincent De Paul Cath 174-A3 NN
Seaford Bapt 125-K1 Y
Seaford Ch of Christ 110-A10 Y
Second Bapt East End 174-D2 NN
Seventh Day Adventist 88-E10 J
Shiloh 56-K7 J
Shiloh Bapt 108-H5; YE-B6 Y
Smith Mem Lightfoot 72-J6 J
South York Bapt 142-D4 Y
Tabernacle 40-B7 NK
Tabernacle United Meth 126-K9 Pq
Temple Beth El 89-E8; WE-B4 W
Temple Sinai 157-C1 NN
Trinity Luth 157-J9 NN
Trinity Meth 143-E1 Pq
Walnut Hills Bapt 89-D9 W
Warwick Mem United Meth 124-D10 NN
Wellspring United Meth 88-K2 J
Wesley Meth 158-K2 H
Williamsburg 89-C9 W
Williamsburg Bapt 89-G7; WE-B3 W
Williamsburg Comm Chapel 88-E10 J
Williamsburg Mennonite 56-H10 J
Williamsburg Presb 89-G7; WE-C3 W
Williamsburg United Meth 89-F7; WE-B4 W
York Assembly of God 109-C10 Y
York Bapt Temple 125-H4 Y
York River Bapt 56-J8 J
Yorkminster Presb 125-E3 Y
Yorktown Bapt 108-J4; YE-D3 Y
Young Kwang Bible Presb 126-J10 Pq
Zion 72-K7 J; 110-A9 Y
Zion Bapt 174-D4 NN
Zion Prospect 125-F9 Y

POINTS OF INTEREST

Abby Aldrich Rockefeller Folk Art Ctr WE-D4 W
Aeromodel Field 124-J3 Y
Air Power Pk & Mus 159-A4 H
Alumni House HE-F6 H
American Artillery Pk 108-J8 Y
Anheuser-Busch Brewery 106-F2 J
B T Washington Mem HE-G6 H
Ballard House YE-D3 Y
Bassett Hall WE-E4 W
Bemis Lab HE-F8 H
Bruton Hgts Sch Educational Ctr 89-J6 W
Buckman Hall HE-G6 H
Carters Grove Mansion 107-A8 J
Cascades Ctr WE-D1 W
Clark HE-F7 H
Coll of William & Mary Muscarelle Mus of Art WE-A4 W
Colonial Natl Hist Park Service Maintenance YE-A6 Y
Colonial Williamsburg Group Arrivals WE-D1 W
Comm HE-F6 H
Customhouse YE-D2 Y
Discovery Ctr 124-E2 NN
Dubois HE-F8 H
Dudley Digges House YE-D2 Y
Eagles Lodge 70-H10 CC
Early Childhood HE-G6 H
Edmund Smith House YE-D3 Y
Emancipation Oak HE-H6 H
Endview Plantation Hist Site 107-K7 NN
First St House 104-F7 J
Fort Eustis Transportation Bldg 123-J4 F
Fort Monroe Chapel of the Centurian 160-A10 FM
Fort Monroe Engineer Wharf 160-A10 FM
Fort Monroe Main Casemate Mus 160-A10 FM
Fort Monroe Old Pt Lighthouse 160-B10 FM
Freeman Hall HE-G6 H
George Washingtons Hdq 108-E8 Y
Glasshouse 104-D5 J
Gymnastics Ctr 140-E7 NN
Hampton Arts Commission HE-B7 H
Hampton Carousel Pk Plaza HE-D5 H
Hampton Comm Services Ctr 159-C6 H
Hampton Roads Naval Mus 175-K9 NBN
Hampton Univ Emancipation Oak HE-G6 H
Hampton Univ Mus 159-F7 H
Harbour Ctr Bldg HE-E5 H
Harkness Hall HE-F8 H
Herbert House 159-F8 H
Historic Hilton Vill 157-F5 NN
Jamestown Settlement 104-E5 J
Japanese Tea House 157-A1 NN
Jaquelin Ambler House 104-G7 J
Jefferson Statue WE-C3 W
Katharine House HE-D7 H
King Hall HE-G6 H
Kittrell Hall HE-G6 H
Lee Hall Mansion Hist Site 107-H8 NN
Marine Science HE-F8 H
Mariners Mus 157-B2 NN
Marl Ravine 57-F7 J
Math HE-F7 H
Military Encampment WE-E3 W
Moore House 109-C5 Y
Music HE-F7 H
Natural Science HE-F7 H
Nelson House YE-D2 Y
Newport News Bapt Home 141-H6 NN
Newport News Cruise Terminal 174-A5 NN
Newsome House Mus 174-E2 NN
Old City-Co Courthouse 89-H7; WE-C4 W
Olin Engineering HE-H6 H
On the Hill Cultural Arts Ctr YE-C2 Y
Pate House YE-D2 Y
Peace Gdns 124-G2 NN
Pen Tran Bus Facility HE-D3 H
Peninsula Fine Arts Ctr 157-B3 NN
Phenix HE-F6 H
Powhatan Cr Canoe Access 104-H2 J
Providence Hall Directors Wing WE-E4 W
Providence Hall Executive Wing WE-E4 W
Providence Hall House WE-E4 W
Providence Hall Tennis Courts WE-E4 W
Radio Tower 71-F5 J
Radio Tower (WMBG) 88-K6 J
Radisson Hotel HE-E5 H
Robert Hunt Shrine 104-E7 J
Sessions House YE-D3 Y
Soccer Fields 140-E6 NN
Somerwell House YE-D2 Y
Stone HE-F7 H
Swan Tavern Group YE-C2 Y
The Ch Tower 104-E7 J
University Mus HE-E8 H
US Army Transportation Mus 123-H4 F
Victory Arch 174-A4 NN
Virginia Advanced Shipbuilding & Carrier Integration 174-A5 NN
Virginia Air & Space Mus 159-E7; HE-D6 H
Virginia Living Mus 141-D9 NN
Virginia Power 109-G6 Y
Virginia War Mus 157-G7 NN
Wallace Collections & Conservation WE-E2 W
Watermens Mus 108-H2; YE-C1 Y
Wheelwright WE-D3 W
Wig Wam HE-F7 H
Williamsburg Anderson Blacksmith WE-D3 W
Williamsburg Armistead Site WE-E3 W
Williamsburg Barraud House WE-D3 W
Williamsburg Blair House WE-C3 W
Williamsburg Bracken Tenement WE-D4 W
Williamsburg Brickyard WE-D3 W
Williamsburg Brush-Everard House WE-D3 W
Williamsburg Bruton Parish Ch WE-D3 W
Williamsburg Cabinetmaker WE-D3 W
Williamsburg Capitol WE-E3 W
Williamsburg Carpenter WE-D3 W
Williamsburg Chownings Tavern WE-D3 W
Williamsburg Chr Retreat 55-D3 J
Williamsburg Christiana Campbells Tavern WE-E3 W
Williamsburg Colonial Garden WE-D4 W
Williamsburg Colonial Post Office WE-D3 W
Williamsburg Cooper WE-D3 W
Williamsburg Courthouse WE-D3 W
Williamsburg Crump House WE-D4 W
Williamsburg Custis Kitchen WE-D4 W
Williamsburg Davidson Shop WE-E3 W
Williamsburg Dewitt Wallace Gallery WE-C4 W
Williamsburg Dickinson Store WE-D3 W
Williamsburg Dubois Grocer WE-D3 W
Williamsburg Galt Apothecary WE-E3 W
Williamsburg Gateway Bldg WE-D3 W
Williamsburg Geddy Foundry WE-D3 W
Williamsburg Geddy House WE-D3 W
Williamsburg Governors Inn WE-B2 W
Williamsburg Governors Palace WE-D3 W
Williamsburg Greenhow Lumber House WE-D3 W
Williamsburg Greenhow Store WE-D3 W
Williamsburg Guardhouse WE-D3 W
Williamsburg Gunsmith WE-E3 W
Williamsburg Harnessmaker & Saddlery WE-C3 W
Williamsburg Inn WE-D4 W
Williamsburg Inn Craft House WE-D4 W
Williamsburg Kings Arms Tavern WE-E3 W
Williamsburg Lightfoot House WE-D4 W
Williamsburg Lodge & Conference Ctr WE-D4 W
Williamsburg Ludwell-Paradise House WE-D3 W
Williamsburg Magazine WE-D3 W
Williamsburg Market Sq WE-D3 W
Williamsburg Market Sq Tavern WE-D3 W
Williamsburg Masonic Lodge WE-D3 W
Williamsburg McKenzie Apothecary WE-D3 W
Williamsburg Milliner WE-E3 W
Williamsburg Orrell House & Kitchen WE-D4 W
Williamsburg Palace Green WE-D3 W
Williamsburg Play Booth Theater WE-E3 W
Williamsburg Powell House WE-E3 W
Williamsburg Prentis Store WE-D3 W
Williamsburg Printer WE-D3 W
Williamsburg Public Gaol WE-E3 W
Williamsburg Public Hosp of 1773 WE-C4 W
Williamsburg Raleigh Tavern WE-E3 W
Williamsburg Randolph House WE-D3 W
Williamsburg Robert Carter House WE-D3 W
Williamsburg Robertsons Windmill WE-D3 W
Williamsburg Secretarys Office WE-E3 W

Williamsburg Shields Tavern WE-E3 W
Williamsburg Shoemaker WE-D3 W
Williamsburg Silversmith WE-E3 W
Williamsburg Soap & Candle Factory 72-F1 J
Williamsburg Stith Shop WE-D3 W
Williamsburg Tarpleys Store WE-E3 W
Williamsburg Taylor House WE-D3 W
Williamsburg Theatre WE-C3 W
Williamsburg Timber Yard WE-D3 W
Williamsburg Timson House WE-C3 W
Williamsburg Tucker House WE-D3 W
Williamsburg Wetherburns Tavern WE-E3 W
Williamsburg Wigmaker WE-E3 W
Williamsburg Winery 105-F3 J
Williamsburg Woodlands Family Resort WE-D1 W
Williamsburg Wythe House WE-D3 W
Winthrop Rockefeller Archaeological Mus 106-K7 J
York River St Pk Observation Tower 57-H6 J
Yorktown Allied Encampment Tour 108-E7 Y
Yorktown Battle Trenches 108-J4; YE-E4 Y
Yorktown Battlefield Tour 108-K5; 109-A6 Y
Yorktown Creative Arts Ctr 108-J2; YE-D2 Y
Yorktown French Artillery Pk 108-F7 Y
Yorktown French Encampment Loop 108-D4 Y
Yorktown Grand French Battery 108-K5; YE-E5 Y
Yorktown Pottery 108-J2; YE-E3 Y
Yorktown Redoubts 9 & 10 109-A4; YE-F4 Y
Yorktown Second Allied Siege Line 108-K4; YE-E4 Y
Yorktown Surrender Field 108-J7 Y
Yorktown Untouched British Redoubt 108-J4; YE-C4 Y
Yorktown Victory Ctr 108-G3 Y
Yorktown Victory Mon 108-K3; YE-E3 Y
Youngs Mill Hist Site 140-H4 NN

POLICE STATIONS

Carybrook Field Office 160-A3 H
Coll of William & Mary 89-G8; WE-C4 W
Doolittle Field Office 159-C3 H
Hampton Police & Public Safety Bldg 159-E6; HE-D4 H
Hampton Police Academy & Fire Training Ctr 159-C4; HE-A1 H

Hampton Police Pistol Range 159-C4 H
James City Law Enforcement Ctr 89-B10 J
Langley AFB Security 143-A10 L
LaSalle Field Office 159-B6 H
Newport News Central Precinct 157-F2 NN
Newport News City Jail 174-B4 NN
Newport News Hdq 174-A4 NN
Newport News North Precinct 124-C10 NN
Poquoson 126-J9 Pq
US Naval Supply Ctr 91-C5 Y
Williamsburg 89-G7; WE-C2 W
York Co Public Safefy Bldg 125-E1 Y

POST OFFICES

Barhamsville 39-F8 NK
Buckroe Beach 160-C3 H
Denbigh 124-E10 NN
Fleet 175-H8 NBN
Fort Eustis Military Reservation 123-H5 F
Fort Monroe 160-A10 FM
Grafton 125-F5 Y
Hampton 158-F8 H
Hampton Univ-Stone 159-F7; HE-F7 H
Hayes 92-K7 G
Hidenwood 141-B10 NN
Lackey 108-B4 Y
Lanexa 54-F3 NK
Lee Hall 107-J10 NN
Lightfoot 72-K6 Y
Naval Weapons Sta 108-B2 Y
Norge 72-G1 J
Old Hampton 159-E6; HE-C4 H
Oyster Pt 141-D7 NN
Parkview 157-K8 NN
Phoebus 159-J8 H
Poquoson 126-H10 Pq
Riverdale 158-J3 H
Seaford 110-A9 Y
Toano 56-B9 J
Veterans Admin 159-F8 H
Warwick 157-E4 NN
Wicomico 92-G4 G
Williamsburg 89-G6; WE-C2 W
Wythe 158-J9 H
Yorktown 108-J2; YE-D1 Y

RAILROAD STATIONS

Lee Hall 107-J9 NN
Newport News (AMTRAK) 157-H7 NN
Williamsburg Transportation Ctr (AMTRAK) 89-G6; WE-C2 W

SCHOOLS

Aberdeen ES 158-F5 H
Abingdon ES 92-H4 G
Alfred Forrest ES 158-B4 H
Armstrong ES 159-B9 H

Armstrong Slater Trade Sch HE-F7 H
Asbury, Francis ES 144-C9 H
Atkins, Mary Chr ES 158-J2 H
AWE Bassette ES 158-K8; 159-A8 H
B C Charles ES 140-E5 NN
B T Washington MS 174-E1 NN
Baker, Clara Byrd ES 88-H10; 104-H1 J
Barron ES 159-E3 H
Bassette, AWE ES 158-K8; 159-A8 H
Benjamin Syms MS 159-H2 H
Berkeley MS 89-B7 W
Bethel HS 158-A2 H
Bethel Manor ES 142-C6 Y
Blair, James MS 89-D4 W
Booker ES 159-G1 H
Bradford Hall Alt 174-F1 H
Briarfield ES 158-B10 NN
Bruton Hgts Sch 89-K1 W
Bruton Hgts Sch Educational Ctr WE-E2 W
Bruton HS 73-J10; 89-K1 Y
Bryan, Jane H ES 159-K6 H
Burbank, Paul ES 158-K2; 159-A2 H
Byrd, Rawls ES 105-C1 J
C Alton Lindsay MS 158-G7 H
Captain John Smith ES 159-H3 H
Carver ES 157-J6 NN
Cary, John B ES 159-K3 H
Charles, B C ES 140-E5 NN
Christopher C Kraft ES 158-C3 H
Clara Byrd Baker ES 88-H10; 104-H1 J
Cooper ES 158-J2 H
Coventry ES 141-K4 Y
Crittenden MS 157-J6 NN
D J Montague ES 88-C1 J
Dare ES 125-G5 Y
Davis, Jefferson MS 158-B4 H
Deer Pk ES 141-D9 NN
Denbigh Early Childhood Ctr 124-D10 NN
Denbigh HS 124-C10 NN
Dozier MS 124-B4 NN
Dunbar-Erwin ES & Achievable Dream ES 174-D4 NN
Dutrow ES 124-B7 NN
Eaton, Thomas MS 158-J3 H
Emmanuel Luth ES 159-B9 H
Enterprise Academy 141-F7 NN
Epes ES 124-B9 NN
Forrest, Alfred ES 158-B4 H
Francis Asbury ES 144-C9 H
Francis Mallory ES 158-C6 H
Gildersleeve MS 140-K9 NN
Gloria Dei Luth 159-J1 H
Grafton Bethel ES 125-H7 Y
Grafton HS 125-E4 Y
Grafton MS 125-E4 Y
Greenwood ES 124-F2 NN
Hampton Chr HS 158-J2 H
Hampton HS 158-J7 H
Hampton Roads Academy 141-C3 NN
Heritage HS 158-B10 NN
Hidenwood ES 140-J10 NN
Hilton ES 157-F6 NN
Hines MS 157-F3 NN
Hudgins, Sara Bonwell Reg Ctr 142-C9 H
Huntington MS 174-D2 NN
Jackson Academy 174-A2 NN

James Blair MS 89-D4 W
James River ES 107-B7 J
Jamestown HS 88-D10; 104-D1 J
Jane H Bryan ES 159-K6 H
Jefferson Davis MS 158-C4 H
Jenkins ES 140-G3 NN
John B Cary ES 159-K3 H
John Marshall ES 174-C4 NN
John Tyler ES 159-C3 H
Jones Magnet MS 160-A3 H
Kecoughtan HS 159-K2 H
Kiln Cr ES 141-B1 NN
Kraft, Christopher C ES 158-C3 H
Lafayette HS 72-H10 J
Langley, Samuel P ES 143-G9 H
Lee Hall 107-J10; 123-J1 NN
Lee, Robert E ES 158-G7 H
Lindsay, C Alton MS 158-G7 H
Machen ES 142-J10 H
MacIntosh ES 124-E7 NN
Magruder ES 174-F3 NN; 90-B8 Y
Magruder MS 90-E9 Y
Mallory, Francis ES 158-C6 H
Marshall, John ES 174-C4 NN
Mary Atkins Chr ES 158-J2 H
Matthew Whaley ES 89-H7; WE-C3 W
Menchville HS 140-E5 NN
Merrimack ES 160-B2 H
Montague, D J ES 88-C1 J
Moton, Robert R ES 159-J6 H
Mount Vernon ES 125-J9 Y
Nelson ES 123-K10 NN
New ES 56-D7 J
New Horizons Tech Ctr 142-D10 H
New Horizons Tech Ctr North Campus 124-F3 NN
Newport News Sch Board Admin Offices 141-A9 NN
Newsome Pk ES 174-C1 NN
Norge ES 72-G2 J
Our Lady of Mount Carmel 157-E3 NN
Palmer ES 141-B3 NN
Paul Burbank ES 158-K2; 159-A2 H
Peninsula Cath HS 141-G9 NN
Phillips ES 159-J1 H
Phoebus HS 159-K5 H
Poquoson ES 127-C10 Pq
Poquoson HS 126-J10 Pq
Poquoson MS 127-B10 Pq
Poquoson Prim 126-J10 Pq
Queens Lake MS 90-C5 Y
Rawls Byrd ES 105-C1 J
Reservoir MS 124-B4 NN
Richneck ES 124-G6 NN
Riverside ES 157-A2 NN
Robert E Lee ES 158-G7 H
Robert R Moton ES 159-J6 H
Saint Andrews Epis 157-F6 NN
Saint Marys Star of the Sea 159-K8 H
Samuel P Langley ES 143-G9 H
Sanford ES 140-C3 NN
Sara Bonwell Hudgins Reg Ctr 142-C9 H
Saunders ES 141-H7 NN
Seaford ES 109-J10; 125-J1 Y
Sedgefield ES 157-H4 NN
Smith, Captain John ES 159-H3 H
South Morrison ES 157-G1 NN
Spratley MS 159-H4 H

Student Ctr HE-G7 H
Syms, Benjamin MS 159-H2 H
Tabb ES 142-A2 Y
Tabb HS 126-A10; 142-A1 Y
Tabb MS 125-H10; 141-H1 Y
Tarrant ES 158-E7 H
Thomas Eaton MS 158-J3 H
Toano MS 56-B10; 72-B1 J
Tucker-Capps ES 157-K5 H
Tyler, John ES 159-C3 H
Victory Acad 92-H4 G
Waller Mill ES 89-F4 Y
Walsingham Academy 89-D9 W
Warwick HS 157-D2 NN
Washington, B T MS 174-E1 NN
Watkins ES 141-G8 NN
Whaley, Matthew ES 89-H7 W
Williamsburg Chr Academy 89-C1 W
Williamsburg Montessori 72-F10 J
Woodside HS 124-F3 NN
Wythe ES 174-H1 H
Yates ES 174-F3 H
York Co Sch Board 125-G4 Y
York HS 108-K8; 109-A8 Y
Yorktown ES 108-K9 Y
Yorktown MS 108-H4; YE-B5 Y

SHOPPING CENTERS

Bay Berry Vill 141-G5 NN
Beaconsdale 141-E10 NN
Belo 90-B7 Y
Brentwood Ctr 157-F3 NN
Bridge Shops 124-E10 NN
Buckroe 160-A4 H
Cedar Valley 73-B5 Y
Coliseum Crossing 158-H3 H
Coliseum Mall 158-G3 H
Coliseum Sq 158-H3 H
Colonial Towne Plaza 72-H4 J
Denbigh Crossing 124-G9 NN
Denbigh Specialty Shops 124-D9 NN
Denbigh Sq 124-E9 NN
Denbigh Vill Centre 124-E10 NN
Drug Emporium Plaza 124-H9 NN
Drug Emporium Shoppes 158-J3 H
Ewell Sta 73-B10 J
Fairway Plaza 141-C1 NN
Farm Fresh 90-B7 Y
Farm Fresh Plaza 92-J3 G
Festival Market Pl 106-E1 J; 106-E1 Y
Francisco Vill 157-G4 NN
Gallery Shops at Lightfoot 72-K5 J
Governors Green 88-H10 J
Grafton 125-F4 Y
Greenwood 158-D5 H
Hampton Mall HE-C5 H
Hampton Plaza 158-B6 H
Hampton Roads Ctr 142-C10; 158-E1 H
Hampton Towne Ctr 158-B1 H
Hampton Towner Centre 142-B10 H
Hampton Woods Plaza 158-B1 H
Hayes Plaza 92-K6 G
Heritage Sq 125-G7 Y
Hidenwood 141-B9 NN

Hilton 157-H6 NN
Ivy Farms 157-F2 NN
James York 90-B7 Y
Kiln Cr 141-G4 Y
Kingsgate Green 89-E4 Y
Lanexa Plaza 54-E3 NK
Langley Sq 159-D4 H
Market Pl 106-D1 J; 106-D1 Y
Merchant Square WE-C3 W
Mercury Plaza 158-F4 H
Merry Oaks 124-A5 NN
Monticello 89-E5; WE-A1 W
Monticello Marketplace 88-J7 J
New Market Plaza 157-J7 NN
Newmarket 157-K7 H
Newmarket Fair 157-K6 H
Newport Crossings 124-E9 NN
Newport Sq 141-G7 NN
Nickerson Plaza 160-A3 H
Norge Crossing 56-G10; 72-G1 J
Old Colony Sq 89-C10 J
Olde Towne Sq 88-K2 J
Oyster Pt Ctr 141-E7 NN
Oyster Pt Plaza 140-J4; 141-D8 NN
Patrick Henry Mall 140-K2; 141-A3 NN
Patriot Plaza Outlet Shops 89-C1 W
Patriot Sq 109-B10 Y
Pavilion Sq 158-A3 H
Poquoson 126-H10; 142-G1 Pq
Prime Outlets 73-B9 J
Queens Plaza 159-C6 H
Richneck Ctr 124-E7 NN
Riverdale 158-J3 H
Sherwood Mall 140-E1 NN
Southampton 159-D9 H
Stoney Bk 124-B6 NN
The Shops at Hampton Harbor 159-H7; HE-G5 H
The Vill Shops at Kingsmill 106-F1 Y
The Williamsburg Outlet Mall 72-J6 J
Tidemill 142-H10 H
Todds Ctr 158-E4 H
Victory Ctr 141-E3 NN
Village Sq 140-H2 NN
Village Sq at Kiln Cr 141-F2 Y
Warwick 157-E4 NN
Warwick Denbigh 124-E10 NN
Warwick Vill 157-E3 NN
Washington Sq 125-G5 Y
Williamsburg 89-E6; WE-A1 W
Williamsburg Crossing 89-B9 J
Williamsburg Pavilion Shops 72-K6 J
Williamsburg Pottery & Factory Outlet Shops 72-J5 J
Willow Oaks 159-H1 H
Woodland Plaza 159-H7 H
Wythe Cr Plaza 126-G10 Pq
Yoder Farms 141-A3 NN
York Sq 125-D3 Y
York Vill 125-C1 Y

STATE, COUNTY & MUNICIPAL FEATURES

Admin Bldg HE-F7 H

Carmel Ctr for Justice HE-D4 H
Charles City Co Transfer Sta 70-B7 CC
City of Newport News Res 124-A3 NN
Courts HE-D5 H
Hampton City Hall 159-F6; HE-D4 H
Hampton Courts HE-D5 H
Hampton Courts Facility 159-E6; HE-D3 H
Hampton Jail HE-D4 H
Hampton Old Courthouse HE-D4 H
Hampton Public Safety Bldg 159-E6; HE-D4 H
Hampton Roads Wastewater Treatment Plant 140-E7 NN
Henrico Co Reg Jail East 39-B1 NK
James City Co Govt Ctr 106-A1 J
James City Co Human Services Ctr 88-K1 J
James City Co Landfill (Transfer Sta) 71-H9; 87-J1 J
National Ctr of St Courts 89-H8; WE-C4 W
Newport News City Hall 174-B4 NN
Newport News Courthouse Law Library 174-A4 NN
Newport News Mun Bldg 174-B4 NN
Old Courthouse HE-D4 H
Poquoson City Hall 142-G1 Pq
Prison Farm 140-F7 NN
Public Works Hdq 126-K10 Pq
Sanifill Landfill 141-K8 H
State Health Dept & Social Services Bldg 125-C3 Y
Virginia DMV 158-A7 H; 124-F7 NN; 90-A5 W
Virginia DOT Williamsburg Residency 89-B5 J
Virginia Marine Resource Commission 174-E6 NN
Williamsburg James City Co Courthouse 89-A6 J
Williamsburg James City Co Health Dept 89-D6 W
Williamsburg Mun Bldg 89-G6; WE-C2 W
York Co Admin Bldg 108-H3; YE-C2 Y
York Co Courthouse 108-J3; YE-C2 Y
York Co Finance Bldg YE-C3 Y
York Co Operations Ctr 109-F10; 125-E1 Y
York Co Waste Management Ctr 125-E1 Y
York-Poquoson Courthouse 108-J3; YE-C3 Y
Yorktown Victory Ctr 108-G3 Y
Yorktown Victory Ctr (Commonwealth of VA) YE-A1 Y

79

NOTES

NOTES

NOTES

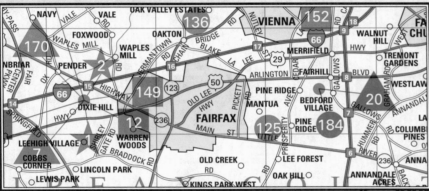

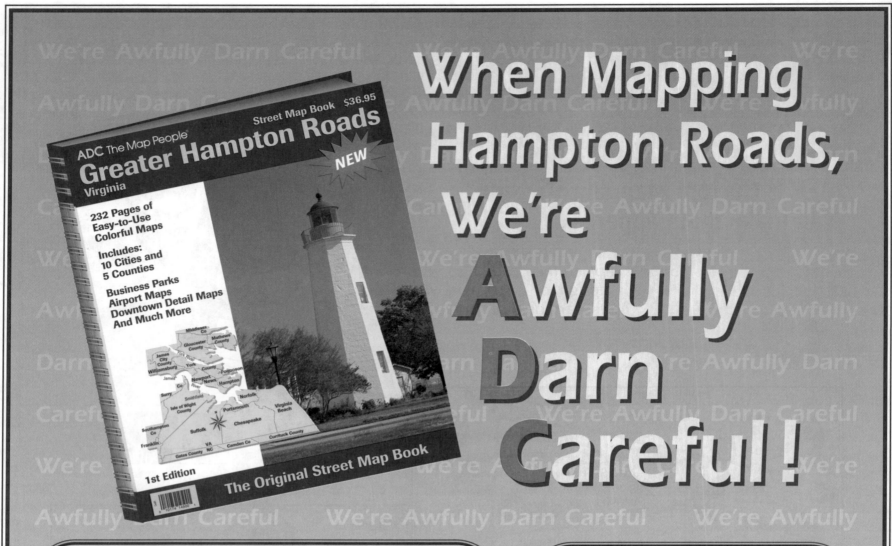